A YEAR WITH

Nature

An Almanac

Marty Crump

Illustrated by Bronwyn McIvor

The University of Chicago Press

Chicago and London

The University of Chicago Press, Chicago 60637
The University of Chicago Press, Ltd., London
© 2018 by The University of Chicago
Illustrations © Bronwyn McIvor

Published 2018
Printed in the United States of America

27 26 25 24 23 22 21 20 19 18   1 2 3 4 5

ISBN-13: 978-0-226-44970-8 (cloth)
ISBN-13: 978-0-226-44984-5 (e-book)
DOI: https://doi.org/10.7208/chicago/9780226449845.001.0001

Library of Congress Cataloging-in-Publication Data

Names: Crump, Martha L., author. | McIvor, Bronwyn, illustrator.
Title: A year with nature : an almanac / Marty Crump ; illustrated by
    Bronwyn McIvor.
Description: Chicago ; London : The University of Chicago Press, 2018. |
    Includes bibliographical references.
Identifiers: LCCN 2018000463 | ISBN 9780226449708 (cloth : alk. paper) |
    ISBN 9780226449845 (e-book)
Subjects: LCSH: Nature—Miscellanea. | Natural history—Miscellanea. |
    Nature conservation—Miscellanea. | Nature—Folklore—Miscella nea. |
    Animals—Miscellanea. | Plants—Miscellanea.
Classification: LCC QH45.5 .C78 2018 | DDC 508—dc23
LC record available at https://lccn.loc.gov/2018000463

♾ This paper meets the requirements of ANSI/NISO Z39.48–1992
(Permanence of Paper).

For Pachamama,

and for Christie

Contents

Preface

The daily entries featured here represent an eclectic assortment of topics and themes relating to our natural world, each appropriate for a specific calendar day. Some were inspired by their introductory quotes, including poetry stanzas and song verses. People are an integral part of nature; thus, I have touched on diverse aspects of human culture, including folklore and spiritual beliefs. I highlight the founding of the National Park Service and the Environmental Protection Agency; publication of classics such as *On the Origin of Species* and *Silent Spring*; musical premieres, including "Swan Lake" and "The New World Symphony"; Beebe's bathysphere trip to the ocean depths; and Neil Armstrong and Buzz Aldrin's landing on the moon. Vignettes include people who have made a difference, from Aristotle to E. O. Wilson; nature expressed in visual arts, music, prose, and poetry; food and animal festivals; the Hope Diamond and La Peregrina pearl; the Chinese zodiac and the Ussher Chronology; citizen science projects; unicorns, the Loch Ness Monster, and flying reindeer; cancer-sniffing dogs and sacred pythons; and discoveries, including the Galápagos Islands, the structure of DNA, and the mountain gorilla. I celebrate *Jurassic Park* and the search for life on Mars; World Chocolate Day and National Potato Day; and the International Day for Biodiversity, World Oceans Day, and International Mountain Day. And more.

I hope these 366 vignettes will cause you to smile, to reminisce, and to contemplate how you can help to ensure that future generations will enjoy the landscapes and rich biodiversity we have been privileged to know and love. Join me on a journey to celebrate the beauty and wonder of our natural world, day by day.

Daily Entries

Wisdom of the Platypus

January 1 is World Day of Peace, a good day to appreciate and respect the differences that characterize our human family. As global
distances shrink and cultures merge, we are becoming a dazzling
kaleidoscope of languages, colors, traditions, and beliefs. Each of
us is special, no one better than another. The Wiradjuri, Australian
Aboriginals from Central New South Wales, tell a story that express-
es this sentiment.

During Dreamtime the Creator made three types of animals:
Mammals, Fish, and Birds. He gave Mammals fur to stay warm on
land; he gave Fish gills to breathe in water; and he gave Birds the
ability to lay eggs out of water. With the leftover bits and pieces, he
made Platypus. After a time, Mammals, Fish, and Birds quarreled,
each claiming to be the best. Mammals asked Platypus, because of
her fur, to join them in their fight against Fish and Birds. Fish encouraged Platypus, because she spends most of her life in water, to

join them in their fight against Mammals and Birds. Birds urged Platypus, because she lays eggs out of water, to join them in their fight against Mammals and Fish. Platypus pondered which group to join. At last she declared, "I am part of each of you; I am part of all of you. I will join no group, and I fight against no group. When the Creator made us, he made us different. We should respect each other's differences and live together peaceably."

Precious Blood

Every two seconds, someone in the United States needs blood. In 1969, President Richard Nixon designated January as National Blood Donor Month. Although an estimated 38 percent of the US population is eligible to donate, fewer than 10 percent of people do so. On average, about 40,000 pints of blood are needed in the country every day. If you can donate, what a way to begin the New Year!

Your donation of a pint can be split into plasma, platelets, and red blood cells, potentially saving three lives. We cannot manufacture or synthesize human blood; it must come from generous donors.

We also depend on the blood from horseshoe crabs. Instead of white blood cells to fight infection, horseshoe crab blood contains amoebocytes that coagulate around bacterial toxins. We use this capability to our advantage. Each year, over 500,000 horseshoe crabs are collected and bled from tissue around their hearts. Their baby blue blood is used to test for contamination in vaccines, injectable drugs, intravenous solutions, and implantable medical devices like pacemakers and knee replacements. The *Limulus* amebocyte lysate (LAL) test is nearly instantaneous: coagulation indicates presence of bacteria; if no coagulation occurs, the solution is considered free of bacteria. One quart of horseshoe crab blood is worth $15,000 in an annual global industry valued at $50 million. Horseshoe crabs can be relieved of one-third of their blood and returned to the ocean. Most people in the US have benefited from horseshoe crab blood, as every vaccine and injectable drug certified by the Food and Drug Administration, as well as surgical implants, must pass the LAL test. We are protected by marine arthropods with an evolutionary history dating back 450 million years.

JANUARY 3

Snake Respect

Like the colorful striped garters eighteenth- and nineteenth-century gentlemen tied just below their knees to hold up their stockings, garter snakes have yellow, tan, or orange stripes running along the length of their long, slender bodies. Garter snakes are abundant in many places in the US. After two elementary school students from Massachusetts learned that their state had no state reptile, they decided the eastern garter snake should have that designation. The students worked with a local representative, and three years

later, on January 3, 2007, then-Governor Mitt Romney declared the eastern garter snake (*Thamnophis sirtalis sirtalis*) the official reptile of Massachusetts.

Snakes engender the extremes of human emotions. We admire and respect them, or we dislike and fear them. And so it is heartwarming for a herpetologist like myself to hear of children who campaigned for a snake to be featured as a state reptile and of a legislature and governor who supported and signed the bill. School children in Massachusetts now learn about the eastern garter snake. Nature centers, museums, and zoos feature this snake in educational exhibits and fieldtrips. If we can empower people with the knowledge to distinguish harmless from venomous species, what to do if they see a snake, and a little about the animals' fascinating biology, most will come to respect the reptiles.

JANUARY 4

From Tadpoles to Ogres

On January 4, 1948, Myanmar (Burma) gained its independence from the United Kingdom. The day marked yet another transformation stage for the Wa, an ethnic group from northern Myanmar who long ago practiced headhunting.

The Wa creation myth tells that as tadpoles, they spent their early years in Nawng Hkeo, a mountain lake at 7000 feet (2134 m) elevation. After they transformed into terrestrial frogs, they lived on a hill called Nam Tao. In time, the frogs became ogres. They lived in a cave; ate deer, wild pigs, goats, and cattle; and were sterile. The Wa searched farther afield for food until one day two ogres encountered humans. They captured and ate a man and returned to their cave with the skull. Eating the man made the Wa fertile, and over time they produced many ogrelets in human form. The parents taught their children that they must keep a man's skull in their settlements to protect against evil spirits and to ensure peace and prosperity.

Tadpoles are a logical focus for a creation myth. Throughout the world frogs symbolize rebirth, reflecting their "magical" transformation from an aquatic, gill-breathing, swimming tadpole into a terrestrial, lung-breathing, leaping frog. Just like frogs, we too pass through critical stages in our lives. Physically, we transition from being bathed in amniotic fluids and getting oxygen through the placenta to breathing air through our lungs once we leave our mother's womb. Emotionally, we experience major changes in our lives and emerge from crises in transformed states.

National Bird Day

Perhaps because birds can do something we cannot—fly, we have long imbued them with supernatural powers. Birds soar in and out of heaven, messengers of the gods. They can foresee the future. They are visible spirits of the deceased, reincarnated human souls.

Birds play major roles in creation myths, as in the sacred story of the Salinan Native Americans of California. Sea Woman, jealous of Eagle because of his grandeur and power, emptied her great basket that contained the oceans and flooded Earth. Eagle gathered the animals together on the only remaining dry land, the peak of Santa Lucia Mountain. He sent Dove to fetch soil, and from this mud Eagle made a new world. Eagle fashioned a man and a woman from elderberry branches and breathed life into them.

Our "feathered friends" raise our spirits. In 2017, Daniel Cox and his colleagues published a paper in *BioScience* that examined the relationship between nature-related experiences of people in their residential neighborhoods and mental-health benefits. They reported that afternoon abundance of birds was positively correlated with a lower prevalence of depression, anxiety, and stress. Birds are many people's favorite animals, perhaps because we see and hear them all around us; we can feed and watch them. Enjoy a walk today, National Bird Day, and see how many of the world's nearly 10,000 species of birds you can spot in your neighborhood.

Pea Pods to Genes

In the late 1940s, folksinger Woody Guthrie exhibited erratic behavior. By 1952, he was diagnosed with Huntington's disease, inherited from his mother. Huntington's disease is a progressive brain disorder that causes uncontrollable body movement and inability to walk, talk, and think. Transmission of the disease follows the laws of Mendelian inheritance.

Austrian monk and scientist Gregor Mendel changed our understanding of inheritance through his experiments on pea plants between 1856 and 1863. Mendel focused on seven traits in his pea plants and found that each trait had two forms, as in yellow seeds and green seeds. He termed the traits dominant and recessive. Mendel concluded that "factors" (now called genes) having these alternate forms result in inheritance of visible traits in predictable ways. Mendelian inheritance refers to simple patterns where inheritance of traits is controlled by genes with two alternate forms.

Huntington's disease is one of these. In everyone's cells, two chromosomes carry the HTT gene: either normal or altered (a gene mutation). Every person who inherits the chromosome with the altered gene will eventually develop Huntington's disease; one copy of the altered gene is all it takes. Thus, a child of an afflicted parent has a 50–50 chance of developing Huntington's disease.

Gregor Mendel, "Father of Modern Genetics," died on January 6, 1884. Evolutionary biology and medicine owe much to Mendel. On an individual level, Mendel empowered us with knowledge to better understand our own inheritance, including tracking and predicting diseases.

JANUARY 7

A New Day, Painted and Framed

Every day a new picture is painted and framed, held up for half
 an hour,
in such lights as the Great Artist chooses, and then withdrawn,
and the curtain falls.
And then the sun goes down, and long the afterglow gives light.
And then the damask curtains glow along the western window.

And now the first star is lit, and I go home.

—HENRY DAVID THOREAU, *Walden*: January 7, 1852

Transcendentalist Henry David Thoreau spent two years, two months, and two days (July 4, 1845, to September 6, 1847) living in a one-room cabin on the shore of Walden Pond, a lake near Concord, Massachusetts. In part he went there to write. More importantly, he sought to "simplify, simplify," to live "deliberately," to immerse himself in nature. He sought to discover what was really important in life, to experience simple living, and to better understand society. Thoreau recorded his thoughts, experiences, and observations in his journal, which later formed the basis for *Walden, or Life in the Woods*.

Thoreau is considered one of the greatest of all American nature writers and is read and respected worldwide. He shares with us the details of nature's happenings—each new picture painted and framed, as he wrote in *Walden* on January 7, 1852. He encourages us to rekindle our wonder and love of nature. Thoreau's reflections have inspired generations of activists, naturalists, conservationists, philosophers, poets, and others to ponder nature, society, friendship, and life itself.

JANUARY 8

Brazilian Adventures

Alfred Russel Wallace, born on January 8, 1823, was one of the greatest natural history explorers of the nineteenth century. Although he formulated the theory of evolution by natural selection independent of Darwin, Wallace focused on biogeography (the study of how organisms are distributed around the world and through geological time) and is primarily known as the "Father of Biogeography."

In April 1848, Wallace left Liverpool on a small trading vessel bound for the Amazon—his first adventure outside of England.

After 29 days at sea, the ship anchored at Pará (Belém), Brazil. Wallace spent over four years exploring the environs of the Amazon River and the Río Negro. In July 1852, he set sail back to England. On day 26 of the voyage, the ship's cargo caught fire. All of Wallace's collected specimens and most of his notes were lost, but he still had his memory. Wallace recorded his Brazilian adventures in *A Narrative of Travels on the Amazon and Rio Negro*, published in 1889.

A Narrative of Travels belonged to a fairly new genre of natural history books during the nineteenth century (others included works by Alexander von Humboldt and Charles Darwin). In addition to vivid descriptions of nature, these books shared personal adventures experienced during exotic travel. Wallace's lively prose has inspired others, including myself, to explore the Amazon Basin. In 1970, my 14-year-old brother, Alan, and I (a second-year graduate student) searched for salamanders along the Amazon River, around Santarém, Óbidos, and Manaus, Brazil, areas that Wallace had explored 120 years earlier. That is what good natural history writing should do—nudge the reader into continued personal discovery.

JANUARY 9

Mermaid—Beautiful or Repulsive?

The earliest known mermaid myth comes from ancient Assyria. The story tells that in about 1000 BCE, the fertility goddess Atargatis accidentally killed her human lover. Out of shame and guilt she dove into the ocean, hoping to become a fish. Her beauty was so great, however, that she retained her goddess-human upper half and only her lower half transformed into a fish.

Mermaids, half-human, half-fish, have been portrayed as both beautiful and repulsive. Beautiful may reflect sailors' wishful thinking from spending months at sea. Repulsive likely reflects sightings of either dugongs or their cousins the manatees, large herbivorous mammals that live in coastal waters off Africa and Australia (du-

gongs) and North America and South America (manatees).

On January 9, 1493, while sailing in the Caribbean, Christopher Columbus wrote that he saw three mermaids rise from the sea. He commented that they were not as beautiful as they have been painted. Columbus's sighting is taken to be the first record by a European of the manatee in the Americas. Mermaid sightings are still reported, most recently off Haifa Bay, Israel, in 2009, and in Zimbabwe in 2012.

If you've ever wanted to be a mermaid, Weeki Wachee Springs near Tampa, Florida, offers mermaid camp. Learn to swim with a sparkly blue, form-fitting tail and execute reverse somersaults as part of the underwater ballet training. We do love fantasy, whether believing or becoming.

JANUARY 10

Incarnated Spirits

Ophiolatry (snake worship) is deeply rooted in Benin (formerly Dahomey) in West Africa, where pythons were believed to control the water supply, Earth's productivity, and human fertility. During the 1700s and 1800s, residents of each Dahomey village cared for a captive python, fed by priests and entertained with song and dance by priestesses. Dahomians worshiped Danh-gbwe, a large python and powerful fetish that served as a go-between for people to approach the divine being. Living pythons were believed to be incarnated spirits of Danh-gbwe. Devotees believed that any child touched by a python had been chosen to become a priest or priestess. Anointed children spent a year in a fetish school to learn the dances and songs of serpent worship.

Snake worship remains alive and well in Benin, especially in the city of Quidah, where pythons are housed in the Temple of Pythons. These snakes are still considered to be incarnated spirits of Danh-gbwe. The temple pythons are set free periodically to roam the city in search of food. Pythons that wander into homes are considered honored guests. In time the snakes are rounded up and returned to the temple.

January 10 is Vodoun Day in Benin, a national, paid holiday to celebrate the Vodoun religion, which boasts a following by about 60 percent of the population. The celebration in Quidah begins with the sacrifice of a goat and continues with chanting and dancing—and honoring of the sacred pythons.

JANUARY 11

What Good Is It?

> The last word in ignorance is the man who says of an animal or plant: What good is it?
> —ALDO LEOPOLD, *Round River: From the Journals of Aldo Leopold*

During the early twentieth century, many conservationists judged the worth of land based on its value to humans, such as abundance of minerals to be mined, density of animals to be hunted, and fish to be caught. Aldo Leopold, professor at the University of Wisconsin, rejected this philosophy and argued that all natural systems and their component plants and animals have worth in their own right. This belief is reflected in the World Charter for Nature, adopted by the United Nations General Assembly in 1982, 34 years after Leopold's death. Signed by well over 100 nations, the charter states: "Every form of life is unique, warranting respect regardless of its worth to man, and, to accord other organisms such recognition, man must be guided by a moral code of action."

Aldo Leopold, born on January 11, 1887, is considered by many to be the father of wildlife management and the US wilderness system. Leopold died in 1948 of a heart attack while helping his neighbors near Baraboo, Wisconsin, fight a grass fire. Not long after his death, his book *A Sand County Almanac* (1949), a collection of essays focusing on land ethics, was published. Leopold's beliefs concerning our relationship to the natural world continue to inspire generations of nature enthusiasts.

JANUARY 12

Gila Monsters—From Persecuted to Protected

Early Spanish explorers to the New World believed that Gila monsters sting with forked tongues and belch toxins. These lizards, which occur from the southwestern United States to northern Sinaloa, Mexico, are in reality less exotic, although they do have venom housed in glands in their lower jaws. Their family name, Helodermatidae, derives from the Greek words *helos*, meaning "the head of a nail or stud," and *derma*, meaning "skin," in reference to their beadlike skin.

The Tohono O'odham from Arizona tell a story of how Gila monsters acquired their beadlike skin of black and pink or peach-colored scales. Long ago all the animals were invited to the first

saguaro cactus wine festival. Everyone arrived in his or her finest attire. Gila monster covered his skin with multicolored pebbles to create a dazzling and durable coat. He enjoyed his festival coat so much that he wears it still today.

During the early 1900s, Gila monsters were displayed as curiosities in roadside shows. Their stuffed bodies were mounted on cactus branches, and their tanned skins were stitched into wallets and belts. By the 1930s, biologists argued that the lizards needed protection. On January 12, 1952, Arizona outlawed collecting or killing Gila monsters, an act that marked the first legal protection of a venomous animal in the United States. Today, Gila monsters receive full legal protection throughout their range in the US and Mexico.

JANUARY 13

Borders Closed

Sometimes we are forced to take extreme measures to protect animals from harm. Such is the case with salamanders in the United

States, home to the world's highest diversity of salamanders at about 190 species (worldwide total is 705 species).

In 2013, Dutch scientists discovered that a fungus, *Batrachochytrium salamandrivorans* (nicknamed *Bsal*), was causing die-offs of the native fire salamanders. The fungus had been carried to the Netherlands on the skin of Asian salamanders imported as pets. *Bsal* has spread elsewhere in Europe, but as of January 1, 2016, there was no record of it in the United States. The US Fish and Wildlife Service (USFWS) estimates that between 2004 and 2014, nearly 2.5 million salamanders were imported into the country, mostly for the pet trade. Inevitably *Bsal* would enter the country, and when it did it could have devastating effects on the native salamanders. The concern was more than national pride in high salamander diversity. Salamanders comprise a large portion of the biomass in some forests, and they play a critical role in food webs and nutrient cycling by eating invertebrates and providing food for many vertebrates.

Biologists decided that the best chance of avoiding major die-offs was to close the US borders to the 201 species of salamanders that could serve as carriers of the deadly fungus. On January 13, 2016, the USFWS published an interim ruling making it illegal to import these 201 species or move them across state lines. As of August 2017, the restriction still holds, and *Bsal* has not yet been discovered in the US. In time we will learn whether this strategy works.

JANUARY 14

Nessie

In 563 CE, in what is now Scotland, the Irish monk Saint Columba began to convert the Picts to Christianity. According to an account written by Adomnán, the Abbot of Iona Abbey, in 565 Columba encountered a group of Picts who were burying a friend by the River Ness. A huge water beast had attacked and drowned the man. Columba sent one of his followers into the water to retrieve the dead man's boat. When the beast approached, Columba made the sign of the cross. The beast fled, and the Picts and Columba's followers gave thanks for the "miracle."

After the "miracle," occasional sightings of a beast were reported from Loch Ness, near the River Ness, but interest in the creature surged in 1933 once a road provided easier access to the lakeshore. On January 14, 1934, Scotland's Secretary of State forbade the capturing or shooting of what had become known as the Loch Ness Monster. There were no laws against photographs, however, and liberties were taken. That same year, a published photograph of "Nessie" turned out to be a toy submarine outfitted with a long neck and small head.

The *New York Times* sponsored a Nessie expedition in the mid-1970s, during which Robert Rines photographed the purported animal. Naturalist Sir Peter Scott and Rines advocated that as an endangered species, Nessie needed protection, and therefore a scientific name. In 1975, they published a paper in *Nature*, bestowing the name *Nessiteras rhombopteryx* ("Ness wonder with a diamond-shaped fin"). Later it was revealed that the scientific name is an anagram for "monster hoax by Sir Peter S." When Scott was accused of chicanery, Rines pointed out that another anagram gives "Yes, both pix are monsters. R."

JANUARY 15

Donkeys and Elephants

American politics is led by donkeys and elephants, and it all began with presidential candidate Andrew Jackson during the election of 1828. Republicans insinuated that Jackson's populist beliefs of "Let the people rule!" would be "like herding a bunch of jackasses into Washington, D.C." Jackson embraced the phrase and used the image of the strong-willed donkey to his advantage on his campaign posters. Jackson won the election and became the first US Democratic president. The symbolism was discarded for decades, until influential cartoonist Thomas Nast popularized the donkey as symbol of the Democratic Party in a cartoon published in *Harper's Weekly* on January 15, 1870.

Not long after this cartoon, Republican Ulysses S. Grant considered a run for a third term as President, a possibility strongly opposed by the *New York Herald*. In 1874, Nast (a Republican and friend of President Grant) published a cartoon in *Harper's Weekly* entitled "Third Term Panic," which mocked the *New York Herald*. The cartoon featured a donkey clothed in a lion's skin, scaring zoo animals representing various interest groups. A wobbly elephant about to fall into the pit of inflation and chaos was labeled "The Republican Vote." By 1880 other cartoonists used the elephant to symbolize the Republican Party, and the rest is political history.

JANUARY 16

Black Patent-Leather Faces

Dian Fossey was born on this day in 1932, destined to become the world's leading authority on mountain gorilla behavior. Fossey graduated from San Jose State College in California with a degree in occupational therapy in 1954, but ever since childhood she had dreamed of walking among the wild animals in Africa. In 1963, she spent her entire savings and took out a bank loan to finance a seven-week trip to Africa. At Olduvai Gorge, Tanzania, Fossey met paleoanthropologist Dr. Louis Leakey—a pivotal moment in her life. Leakey told Fossey about Jane Goodall's work with chimpanzees and touted the value of long-term fieldwork. Later in the trip, after seeing mountain gorillas in the Virungas of the Democratic Republic of the Congo, Fossey wrote in her journal: "A group of about six adult gorillas stared apprehensively back at us through the opening in the wall of vegetation. A phalanx of enormous, half-seen, looming black bodies surmounted by shiny black patent-leather faces with deep-set warm brown eyes."

In March 1966, Fossey attended a lecture by Leakey in Louisville, Kentucky. The following day he offered her the position to head up a long-term field project to study mountain gorillas. In December 1966 Fossey returned to Africa, where she lived with mountain gorillas for most of the next 19 years until she was murdered by an unknown assailant(s) in her cabin in the Virunga Mountains. Fossey is buried at Karisoke, Rwanda, next to her beloved gorilla Digit, in the gorilla graveyard. She gave her life for the animals she loved.

JANUARY 17

No Tree . . . No Me!

An Australian aboriginal legend explains why koalas have no tail. Long ago, during a major drought, the other animals noticed that Koala did not suffer from thirst. They watched him closely, assuming he had a secret source of water. The animals saw nothing until one day Lyrebird watched him climb a tree, hang upside down by his long tail, and sip water that had collected in the fork of branches. Lyrebird guessed the tree was hollow and full of water. He strutted back to camp and returned with a firestick. Lyrebird torched the tree, and as the trunk exploded, water spewed everywhere. He and the other animals quenched their thirsts, at a cost. Fire scorched the outer edge of Lyrebird's tail feathers and turned them brown, while flames devoured Koala's tail.

Although people everywhere adore koalas—their stout bodies and round faces bordered by fluffy ears resemble teddy bears and they seem to plead "Cuddle me!"—their populations are declining because of people. Aboriginals hunted them for food, and during the nineteenth and early twentieth centuries millions were shot for their fur. Now their major threat is habitat fragmentation and modification. An estimated 80 percent of the animals' original habitat has been destroyed. January 17, 2017, marked the 31-year anniversary of the founding of the nonprofit Australian Koala Foundation.

The foundation focuses on research, conservation, and community education. Its slogan reads: "No Tree . . . No Me."

JANUARY 18

Paradise on Earth

Exposed peaks of an underground volcanic mountain range rise up from the deep blue North Pacific Ocean, forming an archipelago of eight major islands, many smaller islets, and several atolls. The landscape and surroundings include rainforest, desert, sand dunes, canyons, active lava flows, waterfalls, white sandy beaches, black sandy beaches, powerful surf waves, coral reefs, and towering cliffs. These, the Hawaiian Islands, are the most isolated group of islands on Earth. On January 18, 1778, British explorer and Royal Navy Captain James Cook and his men were the first Europeans to set foot on the islands. Cook named them the Sandwich Islands in honor of his patron the Earl of Sandwich.

The Hawaiian Islands are home to many endemic plants and animals that evolved in an isolated world, but which have been impacted by colonizing humans. Early Polynesians who settled on the islands believed that gods and spirits inhabited all aspects of nature. Oral tradition says that Native Hawaiians are the living descendants of the Earth Mother and the Sky Father, and that their ancestors include nature deities—forest, ocean, winds, and rains. Yet they, like any people in any ecosystem, modified their surroundings. They introduced pigs and rats, and they cut down forests. Later, Europeans and their domesticated animals further degraded the landscape and caused the extinction of numerous endemics, including 23 species of birds. A wake-up call has resulted in dedicated efforts to preserve the islands' fragile ecosystems, including establishment of national parks and wildlife refuges, restrictions on commercial fishing, and protections against invasive species introductions. For many, the Hawaiian Islands are a slice of paradise on Earth.

Cu Rua

According to a fifteenth-century Vietnamese legend, Golden Turtle God gave Emperor Le Loi a magic sword to drive out the occupying Chinese forces. After the land was free, while Le Loi was boating on a lake, a large, golden turtle surfaced and reclaimed the sword. The turtle's lake in central Hanoi was renamed Hoan Kiem, "Lake of the Returned Sword." This turtle became known as Cu Rua (Great-Grandfather Turtle), a symbol of Vietnamese independence and endurance. In the 1880s, a shrine was built in the lake to honor Cu Rua.

In recent decades, a Yangtze giant softshell turtle, believed to be descended from the legendary turtle and also named Cu Rua, paddled around the lake. On January 19, 2016, the deceased body of Cu Rua was found floating on the lake surface. Death apparently came from natural causes. The *New York Times* reported: "It would be difficult to overstate Cu Rua's spiritual and cultural significance in this deeply superstitious and Confucian country, where the news of the turtle's demise prompted an outpouring of sadness and hand-wringing." The 373-pound (169-kg) turtle was thought to have been 80–100 years old.

Cu Rua's body was placed in cold storage in the Vietnam National-al Museum of Nature. In April 2016, specialists began to plastinate Cu Rua's body (plastination is a preservation technique of replacing water and fat with plastics). The process is expected to take up to 1.5 years to complete. Such is the value placed on the animal symbolic of Hanoi's resistance to Chinese aggression.

Language of Flowers

Just as we have birthstones, we have birth flowers. Folk wisdom tells that the characteristics associated with a birth flower become part of every person born in that month. The idea of birth flower may date back to the Roman Empire, as the ancient Romans are thought to have been the first to give seasonal flowers as birthday gifts. The Victorian Era (1837–1901) was a conservative time, during which etiquette forbade demonstrative expression of love or affection. An extensive and complex "language of flowers" was developed as a means for lovers to exchange romantic messages. Each flower, and in some cases each color of a flower, symbolized an emotion or wish to be conveyed. From this language of flowers, the symbolism associated with birth flowers has evolved.

The birth flower for people born in January is the carnation, a symbol of fascination, love, admiration, innocence, and distinction. The natural (original) color of carnations is pale pink, purplish-pink, or peach; cultivars come in mainly yellow, red, and white. Each color is associated with a particular message. Yellow carnations symbolize disappointment and rejection. Red carnations represent deep love and affection. White carnations symbolize

pure love. Pink carnations reflect a mother's undying love, stemming from a legend that pink carnations first appeared on Earth where Mary's tears fell as Jesus carried his cross. Pink carnations are traditionally given to one's mother on Mother's Day.

Anti-theft Security

In 2001, Christy McKeown, a North Carolina wildlife rehabilitator, announced an unofficial holiday on January 21 to acknowledge the role that squirrels play in our lives—Squirrel Appreciation Day. The day is still celebrated annually at nature centers, natural history museums, and other venues that focus on wildlife. We appreciate squirrels because sometimes they forget where they buried their nuts, and the seeds germinate and grow into forest giants. But beyond that, squirrels fascinate us. Take their thieving and antithieving behavior, for example.

Squirrels dig up other squirrels' caches and help themselves to a free lunch. In fact, squirrels may lose 25 percent of their buried nuts to other squirrels. To protect themselves, eastern gray squirrels pretend to bury nuts to throw off potential thieves that might be watching. This trickster behavior, called "deceptive caching," takes more than one form. In Pennsylvania, *before* gray squirrels bury their nuts, they sometimes hold the nuts between their teeth, dig holes, vigorously re-cover the holes, and repeat the process without burying the nuts. The result is a trail of empty cache sites preceding the buried treasure. In Connecticut, *after* gray squirrels bury nuts they sometimes move to other spots where they dig and re-cover holes without burying nuts, or they just feign covering behavior.

No wonder it's so difficult to design a squirrel-proof birdfeeder! Instead of shooing away the bold squirrel pilfering seeds from your "squirrel-proof" birdfeeder, you could enjoy its acrobatics and admire its determination.

JANUARY 22

Pleasure in the Pathless Woods

> There is a pleasure in the pathless woods,
> There is a rapture on the lonely shore,
> There is society, where none intrudes,
> By the deep sea and music in its roar:
> I love not man the less, but Nature more,
> From these our interviews, in which I steal
> From all I may be, or have been before,
> To mingle with the Universe, and feel
> What I can ne'er express, yet cannot all conceal.
> —LORD BYRON, "Childe Harold's Pilgrimage"

George Gordon Byron (Lord Byron) was born on January 22, 1788, destined to become one of the great British poets and a leading figure in the Romantic Movement. The Romantic Period, from the late 1700s to about 1850, emphasized emotion and appreciation of the beauty and wonder of our natural world over scientific rationalization of nature. Romanticism attempted to reunite people with nature, and it opposed industrialization, viewed as causing people unhappiness because of long working hours and deplorable working conditions. Visual artists, musicians, writers, and intellectuals embraced Romanticism. They supported and encouraged each other, as for example reflected in the pieces composed by Felix Mendelssohn and Robert Schumann who set Lord Byron's poetry to music.

The above stanza reflects Byron's profound bond with his natural surroundings and his craving for solitude to explore and rejoice in nature. Byron was not a misanthrope—he simply enjoyed and appreciated nature more than people. So too, many of us in the twenty-first century find pleasure in the pathless woods where we can feel one with nature without intrusion from society.

JANUARY 23

The Incredible, Edible Cricket

In the fall of 2012, Gabi Lewis, a senior at Brown University in Providence, Rhode Island, struggled to find the perfect protein bar—one that tasted good and was healthful. One day Lewis's roommate, Greg Sewitz, raved about the benefits of insect protein after attending a conference. Aha! Bake crickets into a tasty protein bar!

In January 2013, the roommates ordered 2000 crickets, which they dried and ground into cricket flour. They concocted a recipe for the "perfect" protein bar. The two graduated in May, moved to New York City, and made their dream a reality. By 2016, hundreds of thousands of Exo Protein bars—cacao nut, peanut butter and jelly, blueberry vanilla, and apple cinnamon—had been sold and received rave reviews.

Crickets are a complete protein source; they contain more than

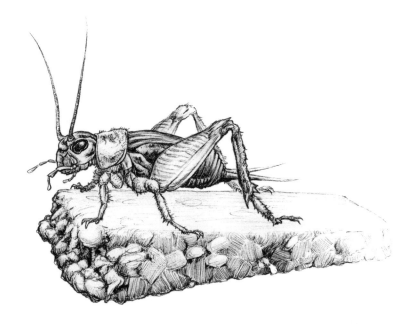

twice as much iron as spinach; and it takes one gallon of water to raise one pound of crickets, compared to nearly 2000 gallons for one pound of beef. In 2014, Kevin Bachhuber founded Big Cricket Farm, the first American insect farm to receive food-grade certification from the Food and Drug Administration. On January 23, 2015, he spoke at the Youngstown, Ohio, TEDx Event and praised the edible cricket. If the idea of crunching an insect and extracting its leg from between your teeth isn't appealing, try cricket flour for high protein smoothies, chocolate chip cookies, and brownies.

JANUARY 24

It's OK to Be Different

January is "It's OK to Be Different Month." The focus is on accepting differences among us—physical and mental abilities, racial and ethnic backgrounds, philosophies and value systems, interests, talents, and appearance. But let's extend the idea to accepting other animals that are very different from us.

We feel the greatest kinship with diurnal animals active in places where we can see them—aboveground or in the air. We admire animals that we consider beautiful, graceful, intelligent, brave, and industrious. We respect those that are helpful to us: insectivores that eat garden pests; birds and butterflies that pollinate our plants; and animals we have domesticated as companions or beasts of burden. We relate to animals that are good parents, monogamous, or altruistic. We especially like young animals with big eyes, large foreheads, and puffy cheeks—baby seals, kittens, and puppies—because they remind us of our own babies.

In contrast, we tend to understand less well, and thus fear and distrust, nocturnal animals and those that live underground or in deep water. We dislike animals that we consider ugly, clumsy, dull, cowardly, and lazy. We are grossed out by animals that perform "disgusting" (though crucial) activities, like carrion-eating turkey vultures and dung-eating beetles. We kill animals that can potentially harm us, such as scorpions and black widow spiders.

Today might be a good day to reevaluate our perceptions. Each living being has a right to exist. It's OK to be different from humans.

JANUARY 25

Vole: Anagram of Love

In 1842, King Frederick William IV of Prussia commissioned Felix Mendelssohn to write music for a production of Shakespeare's play "A Midsummer Night's Dream." For the wedding scene, Mendelssohn wrote the "Wedding March." British monarch Queen Victoria loved Mendelssohn's music. Her eldest daughter, Princess Victoria, had the "Wedding March" played when she married the Crown Prince of Prussia on January 25, 1858. After that, its use became popular, and today Mendelssohn's "Wedding March" is widely used as a wedding recessional.

Marriage generally carries with it the hope for monogamy, yet mammalian monogamous relationships are rare, occurring in only 3–5 percent of species. One outstanding example is the prairie vole, which forms monogamous bonds often lasting for life. Although their lives are short (1–2 years), the little rodents crank out one litter each month. What magic cements the pair bond? Hormones. Experiments have shown that oxytocin encourages the female to huddle with her mate, and vasopressin encourages the male to stay put. Researchers hypothesized that if oxytocin and vasopressin were released during sex, the hormones could foster this strong bond. It turns out that prairie voles have receptors for these hormones in areas of their brains associated with reward and addiction. What this means is that the voles associate the reward of sex with the presence of a particular partner. Males stick around and care for their offspring because of the positive feelings of monogamy.

Not surprisingly the brains of meadow voles, the promiscuous cousins of prairie voles, are wired differently. There are no warm, fuzzy feelings, and the males abandon their mates and young and search for their next partners.

JANUARY 26

Bird of Courage

Imagine choosing your new country's emblem. How should the country project itself to its citizens and to the world? If an animal is to represent the country, should it be gentle like a Zenaida dove (Anguilla)? Strong and powerful like a snow leopard (Afghanistan)?

On January 26, 1784, two years after the bald eagle had been adopted as the emblem of the United States of America, Benjamin Franklin wrote to his daughter that a better choice for America's emblem would have been the turkey. "For my own part I wish the Bald Eagle had not been chosen the Representative of our Country. He is a Bird of bad moral Character. He does not get his Living honestly. For the Truth the Turkey is in Comparison a much more respectable

Bird. He is besides, though a little vain and silly, a Bird of Courage, and would not hesitate to attack a Grenadier of the British Guards who should presume to invade his Farm Yard with a red Coat on."

Franklin knew his animals. Bald eagles steal food from ospreys, though they are perfectly capable of catching their own fish. And anyone who has been around turkeys knows that these feisty birds make great guard animals. It's just as well that the turkey is not our national emblem. If it was, Thanksgiving would likely feature a different entrée.

JANUARY 27

Vicarious Adventure

Most of us don't have the opportunity to study chimpanzees in Tanzania; excavate at Machu Picchu; search for ancient shipwrecks; investigate volcanoes, earthquakes, or glaciers; unearth hominid remains in East Africa; or explore the ocean depths. But thanks to the National Geographic Society, we can experience these adventures vicariously.

In mid-January 1888, 33 scientists, scholars, explorers, and wealthy patrons interested in travel gathered at the Cosmos Club in Washington, DC, to begin the process of founding a society "to increase and diffuse geographic knowledge." Two weeks later, on January 27, the National Geographic Society was founded. Gardner Greene Hubbard, a financier and a founder and first president of Bell Telephone Company, served as the society's first president. His son-in-law Alexander Graham Bell succeeded him as the second president of the National Geographic Society.

To date the National Geographic Society has awarded more than 12,000 grants to scientists and explorers. The society shares the recipients' results with the world, through the *National Geographic* magazine, television programs, books, and other media. For the person who wants more than vicarious adventure, the National Geographic Society sponsors travel expeditions on every continent. The National Geographic Society believes in "the power of science, exploration, education, and storytelling to change the world." Through exquisite photography and unbeatable prose, the National Geographic Society encourages us to care about nature and to protect our planet.

JANUARY 28

Combat Dance

Two prairie rattlesnakes circle each other. They rear upward. The larger snake coils around his opponent, the two resembling a braided rope. They separate, meet chin-to-chin, embrace and separate, again and again. Each tries to force his opponent to the ground to exert his dominance. Finally the larger rattler rises above the other and knocks him down. The loser slithers away. The snakes have executed their combat without striking or intent to kill. Chances are good there is a female nearby. The winner will mate with her, while the loser must search elsewhere.

The National Museum of Animals & Society in Los Angeles, California, recognizes January 28 as Rattlesnake Appreciation Day. Rattlesnakes improve our lives by doing what they do best—eating rodents. Rodents cause millions of dollars worth of damage to agricultural crops and stored grain. They also carry diseases such as hantavirus. Ticks that transmit Lyme disease parasitize rodents, so by eating rodents, rattlesnakes reduce the likelihood that we will contract the disease. In 2013, researchers from the University of Maryland reported that each year a single adult timber rattlesnake can remove up to 4500 of the ticks that cause Lyme disease. Rattlesnake expert Laurence Klauber gives us yet another reason to appreciate rattlesnakes—an aesthetic one: "They are expert performers on a musical instrument which they themselves cannot hear."

JANUARY 29

Buddy

Morris Frank lost the sight of one eye from a childhood accident. He lost sight in the other eye in a boxing match when he was 16. Three years later, in 1927, Frank responded to an article written in the *Saturday Evening Post* by Dorothy Eustis, a dog-breeder and trainer in Switzerland, who told of German shepherds guiding blinded World War I veterans. Frank wrote to Eustis: "Thousands of blind people like me abhor being dependent on others. Help me and I will help them. Train me and I will bring back my dog to show people here how a blind man can be absolutely on his own."

Eustis invited Frank to Switzerland. Regulations at the time required that because Frank could not fend for himself, he would be classified as a "package." Frank traveled by steamship from Nashville, Tennessee, to Switzerland, as an American Express package.

After the training, Frank returned to Tennessee with his German shepherd, Buddy. In 1928, Eustis and three trained German shepherds traveled to Nashville. Eustis and Frank formed The Seeing Eye school for guide dogs, the oldest such school in the United States. Frank traveled widely with Buddy (and subsequent Buddys) to break barriers and create regulation allowing guide dogs to lead the blind into places they needed to go.

January 29 is National Seeing Eye Dog Day. Many visually impaired people have Morris Frank, Dorothy Eustis, and Buddy to thank for their ability to navigate the world independently.

JANUARY 30

Who Taught You?

January is National Mentoring Month in the United States—a good time to ask yourself: "Who taught you to pay attention to nature?" Who shared his or her passion for nature with you and passed on the joy of observing, asking questions, hypothesizing, and seeking answers?

For me, that special person was my father, a hardrock mining geologist. When I was a child, Dad and I took long "nature walks" in the woods behind our house in the Adirondack Mountains of upstate New York. We searched for trillium and pink lady's slippers and marveled over their exquisite beauty, each one prettier than the former. We collected cardinal and blue jay feathers. Dad helped me catch tadpoles and together we watched the magic of leopard frog metamorphosis. He encouraged me to observe and ask questions.

I passed on the love of nature to my daughter. As I write these words, Karen is texting me from her home in Ireland where she and her six-year-old daughter, Fionna, have been incubating five duck eggs. In the wee hours of the morning, they are watching their first duckling pip inside its egg. The excitement is palpable. Mentees become mentors.

Ham

He was born in the summer of 1957 in Cameroon, Africa. After his capture, he was sent to the Rare Bird Farm in Miami, Florida. In 1959, he was purchased by the United States Air Force and shipped to Holloman Air Force Base, New Mexico, where he became known as No. 65. At the base, he competed against 40 others of his own kind, pushing levers in response to electric lights and sounds. No. 65 had impeccable reaction speeds and he was a fast learner, buoyed by banana rewards. He won the competition and earned the right to pave the way to putting an American astronaut in space.

On January 31, 1961, chimpanzee No. 65 was launched into space from Cape Canaveral, Florida. No. 65 performed his assigned tasks in space with flying colors. His lever-pushing performance was only a fraction of a second slower than it had been on land, demonstrat-

ing that motor tasks could be performed effectively in space. The capsule splashed into the Atlantic Ocean and was recovered by rescue ship. No. 65 survived his 16 minutes and 39 seconds to and from space, seemingly calm and in good spirits, with nothing worse than a bruised nose.

Now a hero, No. 65 had earned the right for a name. NASA dubbed him Ham, an acronym for Holloman Aerospace Medical Center, the lab that prepared him for his historic mission. Ham's return suggested that humans could do it also, and on May 5, 1961, astronaut Alan Shepard became the first American to travel into space.

FEBRUARY 1

Feed the Birds

February is a tough month for birds in the Northern Hemisphere. It's cold, food is scarce, and liquid water may be hard to find. In February 1994, John Porter (R-IL) read a resolution into the US Congressional Record recognizing February as National Bird-Feeding Month. The resolution encourages people to provide food, water, and shelter to help wild birds survive the winter.

People have responded to the resolution. Today, more than 40 percent of Americans feed their backyard birds. Some keep it simple and spread mixed seed on a homemade wooden platform, while others fill multiple bird feeders at different heights with a variety of foods to attract their favorite neighborhood birds. Options include black oil sunflower seeds and Nyjer seed for goldfinches, chickadees, nuthatches, cardinals, and house finches; millet for juncos and sparrows; cracked corn for bobwhite quail, ring-necked pheasants, doves, and jays; fruit for orioles, cedar waxwings, and tanagers; nectar for hummingbirds; and suet cakes for woodpeckers and nuthatches.

The National Audubon Society offers tips for backyard bird feeding. Locate feeders to minimize bird collisions with windows and exposure to outdoor cats. Clean feeders to avoid deadly mold. Offer suet cakes only during cool weather; suet can turn rancid in warm temperatures, and fat drips on birds' feathers can damage their natural waterproofing. Feeding the birds brings nature literally to your window.

Weather Prognosticators

Depending on your preference, celebrate today as either Groundhog Day or Hedgehog Day. In the United States, tradition holds that if it is cloudy when the groundhog from Punxsutawney, Pennsylvania, emerges on February 2, spring will come early. If it is sunny, Punxsutawney Phil will see his shadow and return to his burrow. Winter

weather will continue for another six weeks. Groundhog Day has been recognized officially since 1887.

The origin of the groundhog tradition is uncertain. One story tells that the ancient Romans watched the hedgehog's behavior in early February. If he peered out from his den and saw his shadow, he skedaddled back inside. It meant there was a clear moon and thus another six weeks of winter. Conquering Romans carried this belief to northern Europe. More than 1.5 millennia later, German settlers carried the belief to Pennsylvania. Not finding any hedgehogs around, but seeing plenty of groundhogs that superficially resembled the animals they knew, the settlers declared groundhogs their weather prognosticator. Another story suggests that Punxsutawney Phil evolved from the ancient Celtic celebration of Imbolc, an early February holiday that included weather divination. Badger dens and hedgehog burrows were watched to see if their inhabitants were spooked by their shadows when they emerged. Again, if the animals saw their shadows, they retreated inside and spring was a long ways off.

People worldwide look for signs of spring, from opening tree buds to returning robins. Behavior of groundhogs, hedgehogs, and badgers are part of the mix.

FEBRUARY 3

"John of the Mountains"

> Everybody needs beauty as well as bread, places to play in and pray in, where nature may heal and cheer and give strength to body and soul alike.
> —JOHN MUIR, *The Yosemite* (1912)

On February 3, 1998, a 32-cent US stamp was issued with an image of John Muir in Yosemite Valley, California. The inscription read "John Muir, Preservationist."

In 1849, at age 11, John Muir emigrated with his family from Scotland to Wisconsin. He first visited Yosemite in 1868 and fell in love with the mountains. Muir returned a year later and worked as a shepherd, guiding a flock of 2000 sheep to meadows in the High Sierra. While on the journey, he sketched his surroundings and recorded notes on plants and animals. By the 1870s Muir had become a wilderness activist devoted to preserving the landscape he dearly loved. In the late 1880s, Muir used his publications and influence to lobby for the creation of Yosemite National Park. The park was established in October 1890.

In June 1892, Muir and others met in San Francisco, California, to write the articles of incorporation for the Sierra Club, an organization that would help protect the Sierra Nevada Mountains. One week later, the charter members elected Muir the organization's first president, a position he held until his death in 1914. The Sierra Club continues to be a premier environmental preservation organization, a celebrated legacy of "John of the Mountains."

FEBRUARY 4

Stuff'Em

Today is National Stuffed Mushroom Day. Culinary creations include just about any ingredient-flavor combination you can stuff into a mushroom cap: sausage, rice, and Cajun spices; crab, shrimp, and Parmesan; spinach, walnuts, and avocado; chorizo and creamy corn; basil and hummus; zucchini and quinoa; bacon, caramelized onions, and blue cheese.

Mushrooms are the fruiting bodies of fungi, a kingdom of organisms that heal the earth, keep it running, and protect it. Fungi come in four basic types defined by how they get their nourishment: saprophytic, parasitic, mycorrhizal, and endophytic. Saprophytic fungi (e.g., oyster mushrooms, shiitake, and white button mushrooms) decompose dead organic matter. Without them, Earth would be one big garbage dump. Although parasitic fungi (e.g., honey mushrooms) extract nutrients from their host's tissue, they nourish other organisms. A parasitized, rotting tree often supports more biodiversity than does a living one. Mycorrhizal fungi (e.g., chanterelles, porcini, and truffles) form mutually beneficial relationships with plants. Either by weaving into the plant's root cells or by wrapping around the roots, the fungus brings added moisture and nutrients to the plant and in return takes sugars the host produces. Endophytic fungi (e.g., amadou mushrooms) invade a plant's tissues for nourishment. Rather than harming the host, however, they enhance the plant's ability to absorb nutrients and may produce toxins that protect the host from insects, bacteria, and other fungi.

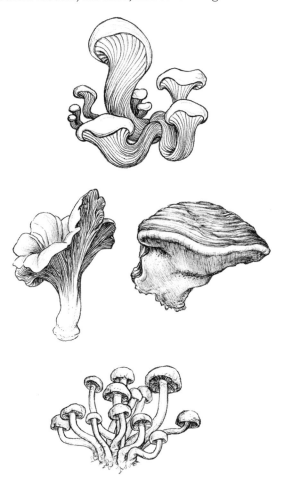

While you enjoy your stuffed mushrooms today, celebrate the benefits that mushrooms offer us.

FEBRUARY 5

To Ease a Broken Heart

If you were born in February, violet is your birth flower. Violets are associated with faithfulness, humility, modesty, purity, hope, and love.

Violet folklore goes way back in time. One Greek legend says that Zeus, king of the gods, fell in love with a beautiful nymph. To protect her from his wife's wrath, Zeus transformed his lover into a white cow. When she wept over her pasture of coarse grass, Zeus turned her tears into violets that only she could eat. A Roman legend tells that one day Venus, goddess of love and beauty, argued with her son Cupid as to who was more beautiful—herself or a group of dancing

maidens. Her son declared the girls' contours lovelier. In a rage, Venus beat the maidens until they turned purple. Cupid then transformed the purple maidens into violets. Both the ancient Greeks and Romans scattered violets around tombs, especially those of children, because of the flower's association with innocence.

The violet has over 200 common names, many of which allude to love: cuddle-me, meet-me-in-the-entry, and kiss-me-at-the-garden-gate. The name flower-of-modesty reflects the flowers' position, partially hidden between heart-shaped leaves. Some common names, such as heartease, reflect the flower's perceived medicinal value. Violets were used to treat heart disease and served as a love potion. Steeped in hot water, the blossoms were believed to ease a broken heart.

FEBRUARY 6

Birds, Bees, and Lame Ducks

> And that's why birds do it, bees do it
> Even educated fleas do it
> Let's do it, let's fall in love.
> —COLE PORTER, "Let's Do It, Let's Fall in Love"

On February 6, 1928, Cole Porter's musical *Paris* opened in Atlantic City. "Let's Misbehave," a favorite song soon renamed "Let's Do It, Let's Fall in Love," features the euphemism "the birds and the bees," referring to sexual intercourse.

Animals appear in many of our everyday expressions. We call others busy bees, greedy pigs, eager beavers, cold fish, cunning serpents, and slippery eels. We refer to swan songs, shaggy dog stories, red herrings, rat races, lone wolves, dark horses, militant hawks, peaceful doves, sacrificial lambs, sacred cows, and white elephants. We pony up, play cat and mouse, go cold turkey, share bear hugs, take the lion's share, wolf down dinner, gather like flies, weasel out

of obligations, horse around, have kittens, make a beeline, chicken out, clam up, and drink like fish. We feel sick as dogs, happy as clams at high tide, and like fish out of water. You can be in the doghouse, have a cow, avoid the elephant in the room, and have ants in your pants. The cat can have your tongue, and we have lame ducks in Congress. Just think how flat our language would be without animal similes and metaphors. Flat as a flounder.

FEBRUARY 7

Timeout and Penalties

If you're not an NFL fan, join the more than two million viewers on Super Bowl Sunday and watch Puppy Bowl. Begun in 2005 on "Animal Planet," the two-hour pre-filmed program is the canine version of football. Two teams of puppies play and wrestle inside a model football stadium. Commentary on the action mirrors football terminology, complete with instant replay shots.

All puppies participating in Puppy Bowl come from shelters, all are adoptable (and adorable), and usually all are adopted by show time. A veterinarian remains on-site during Puppy Bowl, and members of several animal welfare organizations observe and ensure that animal welfare standards are maintained. Timeout is called when puppies get too rough. Penalties are issued for puppies relieving themselves on the field. A touchdown is declared when a puppy drags the football into the end zone. Cameras placed inside toys offer a puppy's-eye view of the action. Peanut butter smeared around the camera lenses allows for additional close-up footage. What puppy can resist a lick?

Partway through, enjoy the Kitty Halftime Show, during which shelter kittens (also adoptable and adorable) play with laser pens and balls of yarn. Catnip strategically placed around the set helps to keep the kittens active. For some viewers, Kitty Halftime outshines the marching bands, pop music, and "wardrobe malfunctions" of Super Bowl.

FEBRUARY 8

Natural Deception

Imagine you're an English naturalist in the Brazilian Amazon rainforest in the mid-1800s. You disturb a large caterpillar perched on a leaf. It stretches and exposes prominent black spots on either side of its body. The spots now resemble reptile eyes. The caterpillar has shape-shifted into a venomous viper. It throws itself backward and slowly sways its viper-like head. You speculate that this behavior might protect a palatable caterpillar from being eaten by looking like an unpalatable animal. You are Henry Walter Bates, the man who wrote the first scientific account of animal mimicry.

Explorer and naturalist Bates was born on February 8, 1825. He was 23 years old when he sailed from Liverpool, England, with his friend Alfred Russel Wallace, into the mouth of the Amazon River where he disembarked at Belém, Brazil. Bates spent the next 11 years (1848–1859) exploring the Amazon Basin. After he returned to England, he wrote many scientific papers and a much-loved book about his tropical adventures: *The Naturalist on the River Amazons.* In one of his most important papers, Bates discussed mimicry among palatable and unpalatable moths and butterflies. He described these mimicries as showing a "palpably intentional likeness which is perfectly staggering." Ecology and evolution students worldwide learn about what we now call Batesian mimicry, and of the man who exposed the natural deception.

The Power of a Dozen Animals

The Chinese New Year celebration begins on the eve of the first day of the first month in the traditional Chinese lunisolar calendar and runs through the 15th of the month. Because a lunar month is about two days shorter than a solar month, every few years an extra month is added to the calendar to account for this difference. Thus, the new year begins on a different day each year and ranges from January 21 to February 20.

Each new year brings the next animal of the Chinese zodiac, a repeating 12-year cycle of a dozen animals: rat, ox, tiger, rabbit, dragon, snake, horse, sheep, monkey, rooster, dog, and pig. Your zodiac sign is determined by your birth year according to the Chinese lunisolar calendar. Birth year is believed to exert a profound influence on personality and destiny because we reflect characteristics of our birth year animal. For example, strengths of people born in Year of the Dog (1946, 1958, 1970, 1982, 1994 . . .) include sincere, responsible, decisive, adaptable, and clever; weaknesses include stubborn and emotional.

The Chinese zodiac, devised more than 2000 years ago for divination and counting the passage of time, is yet another example of how we have incorporated animals into our lives. Belief in the power of animals to influence our personalities can be profound. For example, in 2014, several provinces in China reported a spike in birth rates toward the end of Year of the Horse (associated with an outgoing, energetic and independent personality) to avoid the less favorable Year of the Sheep (associated with a gentle, but somewhat timid nature).

Happy Chinese New Year!

Terrifying Unicorns

> The unicorn . . . has the body of a horse, the head of a stag,
> the feet of an elephant, the tail of a boar, and a single black
> horn three feet long in the middle of its forehead. Its cry is
> a deep bellow.
> —PLINY THE ELDER

In 416 BCE, the physician Ctesias wrote the earliest-known account of the unicorn: a horse-sized wild ass with white body, dark red head, and an 18-inch horn on the forehead. Aristotle (384–322 BCE) believed that unicorns existed, as did Pliny the Elder (23–79 CE), author of *Naturalis Historia*. Accounts in medieval bestiaries depicted unicorns as small horses, goats, or asses with single, straight horns in their foreheads, fierce beasts that could be captured only by virgins.

Fast forward to the twenty-first century. On February 10, 2016, a paper reporting on Siberian unicorns was accepted in the *American Journal of Applied Sciences* (published five days later). A fossil site in Kazakhstan revealed that *Elasmotherium sibiricum*, an extinct genus of rhinoceros thought to have borne a long horn on its forehead, lived 29,000 years ago (measured by radiocarbon dating analysis). Previously, they were thought to have died out over 350,000 years ago. That means that humans and these single-horned animals lived together in northern Central Asia. These Siberian unicorns were not shimmering white with pink or purple manes and silver or gold horns. They were over six feet (1.8 m) tall, weighed four tons, carried weapons on their heads, and likely terrified our ancestors.

FEBRUARY 11

From Author of Jaws to Shark Activist

A great white shark is terrorizing the (fictional) small resort town of Amity on Long Island, New York, savagely dismembering skinny dippers and swimmers just offshore. A professional shark hunter and two other men set out to dispatch the killer shark. Sound familiar? It's *Jaws*, the 1974 novel by Peter Benchley.

Benchley had been working as a struggling freelance journalist. What he longed to do, however, was write a novel about a man-eating shark terrorizing an oceanside resort community. Benchley wrote his novel and eventually signed a contract with Doubleday. The book stayed on the *New York Times* bestseller list for 44 weeks, and Benchley co-wrote the film adaptation for the 1975 movie *Jaws*. Anyone who has seen the movie will remember the ominous music, written by John Williams. Alternating two notes portend approaching danger from the great white shark—raw fear that brings terror and panic.

Later in life, Benchley regretted his sensational portrayal of sharks, which encouraged irrational fear of these keystone predators. His regret peaked when, diving off the coast of Costa Rica in the early 1980s, he encountered shark corpses, minus their dorsal fins, littering the ocean floor—mutilated victims of the worldwide market for shark fin soup. This horrifying discovery converted Benchley into an outspoken advocate and educator for marine conservation. From then on, until his death on February 11, 2006, Benchley fought to protect sharks.

Happy Birthday, Charles Darwin

Charles Darwin, one of the most influential figures in biology, was born in Shrewsbury, England, on February 12, 1809, the fifth of six children. His father was a medical doctor and financier, and his mother was a member of the Wedgwood family. His grandfather, Erasmus Darwin, was a physician and natural philosopher who had written about the heretical idea that species could "transmutate" into other species. As a child, Darwin was an avid collector of insects, minerals, and other natural treasures. At 16 years, he began to study medicine at Edinburgh University but hated anatomy and couldn't stomach surgery. After two years, he quit medicine and attended Christ's College, Cambridge, where he studied divinity, a profession he felt would allow free time for nature study. While at Cambridge, nothing gave Darwin greater pleasure than collecting beetles, as reflected in his autobiography:

> One day, on tearing off some old bark, I saw two rare beetles and seized one in each hand; then I saw a third and new kind, which I could not bear to lose, so that I popped the one which I held in my right hand into my mouth. Alas it ejected some intensely acrid fluid, which burnt my tongue so that I was forced to spit the beetle out, which was lost, as well as the third one. (chapter II)

Darwin became neither a doctor nor a clergyman. In December 1831, he joined an expedition around the world as naturalist on the *HMS Beagle* and spent the rest of his life describing and explaining the natural world.

FEBRUARY 13

The Value of a Yak

Today is Tibetan Independence Day, an expression of Tibetans' desire to be politically separate from the People's Republic of China. In various cities worldwide, groups that support Tibet's independence raise the Tibetan flag and hold events to express their solidarity with Tibetans. When many people think of Tibet, the first images that spring to mind are Mount Everest, the Dalai Lama, the mystical lost valley of Shangri-La, and yaks—shaggy, long-haired members of the cow family.

Tibetans have herded yaks and kept them a focus of their culture for at least 2000 years. They use domesticated yaks for transport, and they weave their traditional robes, blankets, and tents from yak hair. Dried yak dung serves as fuel for fires, and yak-racing is a favorite Tibetan sport. Tibetans eat yak meat, blood, cheese, yogurt, and butter. Yak butter is burned in lamps, and yak butter tea is a high-energy staple of the people's diet.

According to Tibetan tradition, Buddhist monks share with Buddha what they receive from their domestic animals. The first butter made from each female yak is the most precious and therefore should be offered. For the past 400 years, Tibetan monks have made intricate sculptures of flowers, animals, and other figures from colored butter. These sculptures, some taking several months to complete, are believed to create a positive karma that brings harmony into the world.

FEBRUARY 14

Lifelong Love

Why do white plastic doves grace the tops of wedding cakes? They symbolize lifelong love, reflecting the fact that many species mate for life. Other species of doves are monogamous at least during a breeding season. For extraspecial nuptial memories, some couples release trained doves (white homing pigeons). The birds circle the couple and then fly into the sky, symbolizing the newlyweds' departure from their families and the forward journey of their lives together.

This link between doves and love goes back at least to the ancient Greeks who associated doves with Aphrodite, goddess of love. Doves frequently caress each other, thus the supposition of love. Aphrodite was often depicted with a dove or two resting on her hand or shoulder or with doves fluttering around her. Her heavenly golden chariot was said to be drawn by a team of white doves. One myth tells

that Aphrodite hatched from a dove's egg in the sea and was brought ashore by a fish. Greeks identified two manifestations of Aphrodite. One was Aphrodite Ourania, goddess of "pure love" (*ourania* means "heavenly"). The other was Aphrodite Pandemas, goddess of sensual, lustful love and patroness of prostitutes (*pandemas* means "for all the people"). The Greek philosopher Plato expressed these as the two sides of love. The dove symbolized both kinds of love, and both Aphrodite Ourania and Aphrodite Pandemas were represented at weddings.

Happy Valentine's Day!

Ecophobia

On February 15, 1999, educator David Sobel published *Beyond Eco-*

phobia, a book that served as a turning point for nature education. At the time, classroom teachers and environmental educators were teaching second- and third-graders about rainforest destruction, acid rain, and oil spills. Sobel argued that focusing on these disasters for young children could lead to *ecophobia*—a fear of the natural world and its ecological problems. He suggested that if young children cannot understand and control environmental degradation, they might circumvent the pain by distancing themselves from nature. Instead of harping on environmental problems, we should encourage young children to explore nature in their own backyards. In his book, Sobel writes: "What's important is that children have an opportunity to bond with the natural world, to learn to love it and feel comfortable in it, before being asked to heal its wounds."

In 1998, Louise Chawla published a paper in which she reviewed studies seeking to identify the reasons conservationists and environmental educators embrace the environmental values they do. The two most important influences were (1) time spent outdoors in natural habitats in childhood or adolescence and (2) parents, teachers, or other adult role models who fostered their interest in nature. Much less influential was awareness of environmental degradation. Clearly, the best way to encourage future generations to appreciate the need for conservation is to reinforce children's innate fascination with and connection to nature.

FEBRUARY 16

Genes and Mutations

Hugo Marie de Vries, Professor of Botany at the University of Amsterdam, was born on February 16, 1848. Unaware of Gregor Mendel's work, which had been virtually ignored, de Vries carried out genetic experiments with plants in the 1880s and 1890s. He reached the same conclusions regarding inheritance as did Mendel, and published his work in 1900. He was one of three scientists, the others being botanists Carl Correns and Erich von Tschermak-Seysenegg, who independently "rediscovered" the laws of inheritance in 1900.

As a child, de Vries loved plants; later he studied botany at Leiden University. There, he became excited about experimental botany and intrigued with Charles Darwin's theory of evolution by natural selection. In 1889, de Vries hypothesized that hereditary traits were passed from one generation to the next via particles he called *pangenes*. Part of his claim to fame is this discovery, which we now call genes. De Vries was interested in how pangenes caused species to change. He noticed that wild forms of evening primrose differed greatly from cultivated varieties, and that new forms appeared randomly. De Vries' second major contribution was his idea of mutation theory. He speculated that species can evolve from other species through sudden changes in traits, which he called mutations. We now know that mutations provide the raw material—heritable variation—on which natural selection works. De Vries' ideas revolutionized twentieth-century genetics and expanded our thinking about evolutionary processes.

FEBRUARY 17

Delicious, Captivating, and Indispensible

February 17 is Crab Appreciation Day. The first thing many of us think about regarding crabs is food: hot crab dip; crab chowder, gumbo, and bisque; deviled crab; steamed blue crab; crab cakes; crab quiche; crab Newburg; crab risotto, lasagna, and manicotti; fried soft-shelled crab; and crab curry. What else is there to appreciate about these crustaceans?

For starters, crab behavior is captivating. Hermit crabs climb into empty gastropod shells for protection; they nudge sea anemones onto their shells, using the anemones' stinging cells for further protection. Terrestrial hermit crabs sometimes have a hard time

finding larger empty shells when they need to size up. No problem. They yank resident crabs from coveted shells and steal their homes. Male fiddler crabs defend their burrows by using their huge "major claws" in ritual combat. If a male loses his major claw, the "minor claw" begins to grow, while a small claw regenerates in place of the missing major. Jamaican tree crabs live in small pockets of rainwater that accumulate in arboreal bromeliads. Mother tree crabs feed their offspring bits of millipedes and damselfly larvae, remove debris from the nursery, and circulate the water to increase oxygen content. They deposit empty snail shells in the water, which buffer pH levels and add calcium.

Beyond that, many species are scavengers. A crabless world would not be a pretty sight and would dampen our fun at the beach. No wonder the National Museum of Animals & Society in Los Angeles, California, encourages us to appreciate crabs.

FEBRUARY 18

Backyard Bird Count

An estimated 60 million people in the United States birdwatch. Some record the birds that come to their backyard feeders; many others compile life lists of the birds they see. If you enjoy watching birds, the Great Backyard Bird Count might be for you and it's happening about now—the second Friday in February through the following Monday. The event was launched by the Cornell Lab of Ornithology and the National Audubon Society in 1998 and takes place worldwide.

The Count is simple and fun. During the 4-day event, choose a site (anywhere, from your backyard to an exotic vacation spot), watch, and record the number of individuals of each species of bird you see within at least 15 minutes. Do this once or as often as desired, for as long and in as many places as you want. Submit a separate checklist for each day, for each location, or for the same location if you counted birds at a different time of day (see the Great Backyard Bird Count website: http://gbbc.birdcount.org).

While you're having fun, you're also providing valuable information for scientists studying patterns of bird populations. In 2017, participants in 149 countries counted 6,223 species of birds and a total of 29,613,979 individuals. The Great Backyard Bird Count invites you to join the more than 210,000 participants worldwide in the annual event.

FEBRUARY 19

Gonads, a Culinary Delight

February or March sees the annual Sea Urchin Festival in the Italian coastal town of Alghero, in northwestern Sardinia, a festival that attracts seafood lovers from across the world. The exact timing of the festival depends on climatic conditions, to ensure the sea urchins are at their best—slightly sweet, slightly salty.

The edible parts of sea urchins are the gonads. Although the delicacy is often referred to as roe (eggs), it includes both male and female gonads. The gonads are bright yellow, orange, or red depending on the species. Sea urchins have five of these brightly colored, delectable organs. People eat sea urchin gonads in different ways in different parts of the world. In Japan, *uni* are eaten raw as sashimi or in sushi with soy sauce and wasabi. In the Mediterranean, *ricci* (Italy) or *oursin* (France) are eaten raw with lemon juice, or just barely cooked in olive oil and butter and added to risotto, spaghetti, pizza, omelets, and fish soup. The raw *erizos* I ate in Chile were served in a ceviche with chopped onions, cilantro, olive oil, and lemon juice. I found sea urchin gonads to have a smooth, buttery texture, and they almost melt in your mouth, like brie or blue cheese. Like raw oysters, the gonads have that unmistakable, intense "straight from the sea" flavor. If you're an epicure, this Sardinian festival might be one for your bucket list.

FEBRUARY 20

Dinner Dances

Every evening at 5:00 when I headed toward the dog food bin, our long-haired dachshund Conan pranced and spun around in a dinner dance. Before his second year, Conan had survived both a life-threatening bout of pancreatitis and surgery to insert a metal pin to extend a leg bone that had stopped growing. During middle age he nearly slipped a disc. Prednisone resolved that misery, but jumping off the bed was now verboten. Conan was a trooper.

For over a decade, the high point of Conan's day was hiking on the San Francisco Peaks in Flagstaff, Arizona. During summers, he seemingly defied gravity on our 12-mile hikes in Silverton, Colorado. There, the high point was chasing marmots until they disappeared

down their burrows—appropriate enjoyment for a dog bred to chase and flush badgers from burrows. Conan lived life with gusto.

Conan spent his last four years in Logan, Utah. Instead of long hikes, he explored the backyard and ate garden peppers and fallen apples. During his last year, he needed to be carried down steps, but he always climbed back up. Despite being nearly deaf and having cataract-clouded vision, Conan never complained.

Our beloved pet died peacefully in his lair, on the carpet, with my husband and me present. Conan was three months shy of 17 years. Every February 20, Love Your Pet Day, I will raise a glass to Conan's memory in honor of my faithful companion who shared the joys and sorrows of a quarter of my life's journey thus far.

FEBRUARY 21

Backyard Bird Nests

Are you thrilled when birds build a nest outside your kitchen window? What if you could attract more nesting birds to your yard? You can, by setting out nest boxes.

The UK has made the activity an annual event. Every year on February 14–21, the UK celebrates National Nest Box Week, organized by the British Trust for Ornithology in 1997. As forest and other natural habitat is lost, birds are left with fewer nesting sites. The goal of the event is to encourage people to provide nest boxes in their backyards, both to help out the birds that already frequent local areas and to encourage more birds to nest in backyards. Thanks in large part to National Nest Box Week, an estimated five to six million nest boxes across the UK welcome nesting birds.

The Cornell Lab of Ornithology in Ithaca, New York, maintains an interactive website with downloadable nest box and platform plans for a range of North American birds, from the pygmy nuthatch to the Canada goose. You'll also find information on geographical nesting range, nesting season, and nesting habitat for many species. You can either buy readymade boxes, or create your own. Birds everywhere gain from supplemented nesting sites, and you'll enjoy the thrill of watching bird parents bring food for their chicks and of watching those chicks fledge.

FEBRUARY 22

The Buffalo Nickel

During the early 1900s, a campaign was initiated to beautify US coinage. Sculptor James E. Fraser was chosen to design a new nickel. Fraser wanted to create a coin that would be truly American, thus his choice of the bison—a symbol of the Wild West—on one side and the profile of a Native American on the other. The coins, often called buffalo nickels, were minted between 1913 and 1938. Fraser received $3166.15 as payment for his design. The two most likely models for the bison were either Bronx, leader of the bison herd at the Bronx Zoo, or Black Diamond, a resident of the Central Park Zoo.

On George Washington's birthday, February 22, 1913, President William Taft presided at groundbreaking ceremonies for the National American Indian Memorial on Staten Island, New York. There, the first 40 buffalo nickels were distributed, mostly to the 32 Native American chiefs attending the event. The memorial was never built, because of lack of funding combined with the advent of World War I in 1914. Now, more than a century later, a Native American group, the Red Storm Drum & Dance Troupe, is urging President Donald Trump to revive the project.

Check your wallet to see if you have any buffalo nickels. If you don't want it, someone else will. Value depends on year, condition, and mint mark, but coins in good condition range from about $0.40 to $400. More importantly, collectors value this coin for the American images it honors.

FEBRUARY 23

Cavorting Bunnies

According to legend, several Iroquois hunters heard a loud thump and saw a huge rabbit. Soon hundreds of rabbits appeared. The big rabbit thumped a rhythm with his foot, and the other rabbits danced in and out of a circle around him. After a bit, the big rabbit thumped twice, and the rabbits followed each other single file and disappeared up a trail. The hunters were so captivated by the performance that they ran back to their village and relayed the sight to their Clan Mother. She told the hunters that the rabbits had shown them the dance as a way for the Iroquois to express their gratitude for everything rabbits give to them. Ever since, the Iroquois have performed the traditional Rabbit Dance, to give thanks to the rabbit people.

Rabbits are still dancing today. They explode into "bunny binkies" (hop-spins and kicks) when they're "happy." In part because they're so much fun to watch, rabbits are the third most pop-

ular pet in the United States and the UK, after cats and dogs. In the US, February is Adopt-a-Rescued-Rabbit Month. Take your pick: American fuzzy lop, French angora, Flemish giant, Netherland dwarf, or any of the other nearly 60 breeds of rabbits. Rabbits don't require daily walks; they can be litter-box trained; and they're snuggly. At the end of a hard day, pour out your troubles and frustrations to the bunny and let her binkies cheer you.

FEBRUARY 24

The Man Who Loved Plants

Joseph Banks, naturalist and botanist, was born in London on February 24, 1743. As a child, he roamed the Lincolnshire countryside, fascinated with the local plants. Banks studied natural history at Christ Church, University of Oxford, but after his father died in 1761, he inherited the family estate in Lincolnshire and never graduated from Oxford.

Banks remained fascinated with plants, however, and became one of England's most celebrated eighteenth-century botanists. He will long be remembered for encouraging collaboration across national borders. By the late 1760s, Banks was corresponding with Carl Linneaus, the Swedish botanist who formalized the system of binomial nomenclature. Banks advised King George III on London's Kew Gardens and encouraged the King to support voyages to collect plant specimens. Banks' efforts helped catapult Kew to the world's preeminent botanical garden. He joined Captain James Cook's first scientific expedition to the South Pacific, from 1768 to 1771. He met the Prussian naturalist Alexander von Humboldt in 1790, just before Humboldt left for his 5-year exploration of northern South America. The two arranged for Humboldt's specimens, if seized by the British, to be sent to Banks. Later, the men shared and exchanged botanical specimens. In 1795, Banks was invested as Knight Commander of the Order of the Bath. Sir Joseph Banks is honored throughout the world, from geographical features (e.g., Banks Islands in the Northwest Territories, Canada) to plants bearing his name (e.g., the genus *Banksia*, containing about 170 species).

FEBRUARY 25

Shark Lady

Eugenie Clark became fascinated with sharks as a nine-year-old visiting the New York Aquarium. As she gazed into the tanks, she imagined what it would be like to swim with sharks. Later, she spent much of her life doing just that. When Clark began her studies in the 1950s, sharks were considered nothing more than dumb and deadly eating machines. Through her research over the next six decades, she proved them otherwise.

Clark, famed marine biologist (also known as "Shark Lady"), died on February 25, 2015, at 92 years old. She had pioneered the use of scuba gear, made more than 70 deep dives in submersibles, developed the first "test-tube fish babies," and discovered the first effective shark repellent—naturally occurring secretions from a flatfish in the Red Sea. She trained sharks to push targets, and in 1965 gave a trained shark to then Crown Prince Akihito of Japan, later Emperor of Japan, when she visited him. Two years later, he visited her in Florida, and she taught him how to snorkel. (Akihito himself is a noted ichthyologist.) Perhaps Clark's greatest legacy is her connection with the public. She worked to dispel our fear of sharks, shared the wonders of sharks through her books, television specials, and lectures, and promoted marine conservation. She served as a role model for a generation of women marine biologists. Clark made her last dive in her tenth decade, remaining fascinated by underwater life until the end.

FEBRUARY 26

Saved by the Dog

In April 1814, as part of the Treaty of Fontainebleau, the European powers forced Napoleon Bonaparte to abdicate as Emperor of France. They banished him to the Mediterranean island of Elba. In the darkness of night on February 26, 1815, Napoleon escaped. When rough seas knocked him overboard, a fisherman's Newfoundland dog jumped into the water and rescued the escapee. Napoleon returned to Paris and reclaimed the throne—until 100 days later when he was defeated at the Battle of Waterloo and banished by the British to the remote South Atlantic Ocean island of St. Helena, where he lived in exile until his death in 1821.

Newfoundlands, or "Newfies," were originally bred in Newfoundland as working dogs for fishermen, where they pulled both heavy nets and floundering fishermen from the water. Newfoundlands have large, webbed feet, oily, water-resistant coats, and natural swimming ability. These huge dogs (males can reach 175 pounds/79 kg) are known for their intelligence, gentle disposition, strength, endurance, and loyalty.

Although considerable training goes into preparing these dogs for official lifeguard duty, Newfies instinctively seem to recognize when people are in danger of drowning, and they want to help, as in the case of Napoleon. In 1996, the Newfoundland Club of America awarded Boo, a 10-month Newfie, a medal for heroism. Boo and his owner were hiking along the Yuba River in northern California, when Boo spotted a man flailing in the rapid current. Boo dove in and pulled the man, a deaf person who could not speak, to shore. Boo had never been trained in water rescue. He just did what came naturally.

FEBRUARY 27

International Polar Bear Day

In 2008, the polar bear became the first animal to be listed under the Endangered Species Act as Threatened because of global warming. Polar bears' critical sea ice habitat is melting. In 2016, the International Union for the Conservation of Nature and Natural Resources (IUCN) estimated that the global population of polar bears is somewhere between 20,000 and 31,000. By 2050, there may be only one-third that number remaining.

The life of a polar bear begins and ends at the edge of ice, from where it preys on ringed seals. In the spring of 2016, the average atmospheric level of carbon dioxide (the primary "greenhouse gas" that traps heat in the Earth's atmosphere) reached a record high to that date. With ice freeze-up happening later in the year and ice break-up arriving sooner, the bears have less time to hunt ringed seals and store fat reserves. Some are finding alternate sources of food—caribou, snow geese, goose eggs, and even starfish. But over the long term, will goose eggs and starfish sustain these large bears,

which can weigh over 1000 pounds (455 kg)?

If our grandchildren are to live in a world with polar bears, each of us must reduce our carbon footprint. February 27 is International Polar Bear Day, and a good day for each of us to vow to do our part to reduce global warming by decreasing our daily energy use. Reduced energy use means less dependence on the fossil fuels that raise the level of carbon dioxide in the atmosphere and contribute to global warming.

FEBRUARY 28

On the Shoulders of Giants

DNA (deoxyribonucleic acid) carries the genetic instructions that make each living species unique. The molecule was discovered in

1869 by the Swiss chemist Friedrich Miescher, but it wasn't until many decades later that we began to understand the biological function of DNA. That didn't happen until we understood its molecular structure. In the early 1950s, English chemist Rosalind Franklin worked on x-ray diffraction images of DNA and provided the foundation for discovering the structure of DNA. On February 28, 1953, American molecular biologist James D. Watson and British molecular biologist Francis H. C. Crick announced the three-dimensional structure of DNA—a double helix. For their discovery, Watson and Crick, along with New Zealand–born British molecular biologist Maurice Wilkens, were awarded the Nobel Prize in Physiology or Medicine in 1962.

The double helix consists of two strands that wind around each other like a twisted ladder. Each strand is formed from a backbone of alternating sugar and phosphate groups. Each complete twist in the DNA helix consists of ten "rungs," each made up of a pair of nucleobases (either adenine and thymine, or cytosine and guanine) joined by hydrogen bonds. The sequence of bases on each DNA strand determines the code carrying the genetic instructions.

Watson and Crick built on the research of many others, and their discovery was just a beginning. Other scientists figured out how we get from a strand of DNA to a protein. Discoveries are built on other discoveries, as phrased by Isaac Newton in a letter to Robert Hooke in 1676: "If I have seen further, it is by standing on the shoulders of giants."

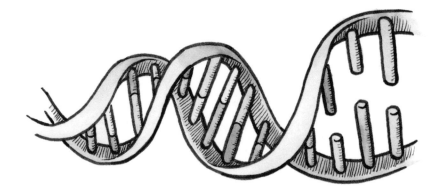

FEBRUARY 29

Supernatural Power and Lunar Eclipses

After Christopher Columbus and his men landed on Jamaica, the indigenous Arawaks welcomed and fed the intruders . . . until the sailors cheated and stole. Desperate for provisions, Columbus consulted his almanac of astronomical tables and noted a forthcoming lunar eclipse. He informed the Arawaks that his God was angry with them for withholding food. In three days God would show his displeasure on the face of the moon. The rising full moon would be nearly obliterated, "inflamed with wrath," foreshadowing the misery that would be inflicted on the Arawaks. On February 29, 1504, the moon began to disappear. The Arawaks panicked, showered Columbus with food, and begged him to intercede with his God. Columbus retreated to his cabin. Once the moon reappeared, he announced that God had forgiven the Arawaks.

Like Columbus, the toad is associated with both supernatural power and lunar eclipses. Instead of a man in the moon, the Chinese see a toad. According to legend, nine false suns once threatened to scorch Earth. Using a magic bow, an archer named Shen I shot the nine false suns out of the sky. The gods rewarded the archer with the Pill of Immortality. Heng O, the archer's wife, stole and swallowed the pill. To escape her husband's wrath, she fled to the moon, where she was transformed into the three-legged toad Ch'an Chu. Shen I dearly loved his wife and built her a palace on the moon. Once each month, the archer visits his wife in her moon palace, explaining the exceptional brilliance of the moon that day. Ch'an Chu, still a troublemaker, occasionally tries to swallow the moon—and causes an eclipse.

MARCH 1

Pet or Meal?

The sacred animal for Ceres, Roman goddess of agriculture and grain, was the pig, perhaps because pigs are so effective at turning over the earth. Romans offered pig entrails to Ceres before sowing their crops, and they sacrificed pigs during harvest festivals to give thanks.

Domesticated pigs are still valued both dead and alive. March 1 is National Pig Day in the United States, celebrated since 1972. Some festivals feature pigs parading around town squares with ribbons entwined around their tails; the intent is to give pigs a break for a day, so street fare consists of beef or chicken hot dogs and portabella sliders. Baby pigs welcome kids at the Children's Zoo in Central Park, New York. Other National Pig Day festivals celebrate the versatility and flavor of pork and offer spare ribs, pork rinds, crispy pork snout, and smoked pig tails. Pig Day in Allentown, Pennsylvania, offers free chocolate-covered bacon for the first 500 ticket purchasers. Despite the fact that Judaism, Islam, and some

Christian denominations forbid the eating of pig, pork is the most commonly consumed meat worldwide, accounting for 38 percent of meat production.

Some people just love pigs. Pot-bellied pigs made popular pets in the 1990s. Micro pigs are all the rage now. Celebrities David Beckham (former English professional football player), Rupert Grint (actor who played Ron Weasley in the Harry Potter film series), and other high-profile personalities have jumped on the porcine bandwagon, touting the virtues of the Basset hound-sized pet pig. Sir Winston Churchill once quipped "I am fond of pigs. Dogs look up to us. Cats look down at us. Pigs treat us as equals."

MARCH 2

Death and Wisdom

> Yesterday the bird of night did sit,
> Even at noonday, upon the market-place
> Hooting and shrieking.
> —WILLIAM SHAKESPEARE, *Julius Caesar*

Owls, about 225 species strong, have long been viewed as symbols of death. Fear of owls likely dates back to our beginnings. Imagine being an early *Homo sapiens*. Nighttime is intimidating, charged with fear of the unknown. Imagine hearing piercing shrieks in the darkness, and wondering when the creature will find and devour you. You cannot identify the shrieks, as you cannot see in the dark. Later in our history, owls were identified as the eerie callers and assumed to be harbingers of death. During the Elizabethan era, the owl, frequently heard from trees near graveyards, was still associated with death. Thus, not surprisingly, William Shakespeare used the owl to portend impending death for the Roman emperor.

We also have associated owls with great wisdom, going back at least to the ancient Greeks. Athena, originally the Greek goddess of the night, later became goddess of wisdom. An owl sat by her side so she could see "the whole truth." Perhaps we associate owls both with death and wisdom because we see them as supernatural. They can do something we cannot: see in the dark.

Join the festivities in Houston, Minnesota, host of an annual International Festival of Owls during the first weekend of March. The mission of the festival is to dispel myths about these nocturnal birds of prey. Events include lectures, nest box building, and dissection of owl pellets (masses of regurgitated, undigested food).

MARCH 3

World Wildlife Day

On March 3, 1973, the Convention on International Trade in Endangered Species of Wild Fauna and Flora (CITES) was adopted. This international treaty protects wildlife and plants by declaring it illegal to sell species listed as endangered, whether live or dead, to another country. It also protects species listed as potentially endangered by limiting trade; these species can be collected and sold only with proper permits. Importantly, consumer nations agree to share responsibility for trade in plants and animals with producer nations. As of August 2017, 183 countries have joined the convention and signed the treaty. CITES is one of the world's most powerful tools to ensure that international trade of endangered plants and animals does not further threaten their existence. The 2017 CITES regulations protect more than 35,000 species of plants and animals.

In December 2013, the United Nations General Assembly proclaimed March 3 as World Wildlife Day, a day to celebrate CITES and to raise awareness of endangered plants and animals. Each year has a different theme. For 2016, the theme was "The future of wildlife is in our hands," reinforcing the philosophy that each of us bears responsibility to safeguard wildlife so that the next generation can appreciate the biodiversity we now enjoy. You can do your part by not buying products made from endangered species, including carved elephant ivory, tortoiseshell, shark fin soup, and traditional Chinese medicines made with parts of lions, tigers, and bears.

MARCH 4

United in Death

Tchaikovsky's "Swan Lake" premiered on March 4, 1877, in Moscow. The debut performance was not well received, in part because of poor production, but critics also proclaimed the music too complex. Times change. "Swan Lake" is now one of the world's most-performed ballets.

The story's origin likely stems from a German folktale, perhaps influenced by a Russian tale. One day an evil sorcerer transforms a beautiful girl named Odette into a swan. She is swan by day and human by night. Odette can regain her full human form only through eternal love from a man who remains faithful to her. In time, Odette and young prince Siegfried fall in love. The sorcerer interferes, however, and disguises his daughter Odile to mirror Odette. Siegfried is tricked. When he realizes what has happened, he begs Odette for her forgiveness. She forgives him, but now their love can never be realized, because she will be a swan forever. Rather than be parted, Odette and Siegfried throw themselves into a lake, to be united in death.

Folklore links swans with true love because in real life, the birds form long-term pair bonds. Swans frequently appear as maidens under enchantment, as their grace, beauty, serenity, and snow-white feathers lend them an otherworldly countenance. But because swans also convey strength and power, some folktales tell of princes and male heroes transformed into swans by evil sorcerers, as in Hans Christian Andersen's *The Wild Swans*. Either way, the stories inevitably involve love and fidelity.

End of an Era

On March 5, 2015, Ringling Bros. and Barnum and Bailey Circus (RBBX) announced that by 2018 they would no longer feature elephant acts. For 145 years, elephants had symbolized the American circus, but protest from animal rights activists and shifting public opinion prompted the phase-out. Elephants live in tight-knit matriarchal family units of typically 6 to 20 individuals. Females spend their entire lives in these groups; males remain with the herd for 12 to 15 years. Elephants are highly intelligent, emotional, altruistic, and empathetic, and so it is not surprising that circus-performing elephants often exhibit aggression and signs of depression. Their spirits seem broken. They may be pining for family members.

The end of prancing pachyderms for RBBX arrived earlier than 2018. May 1, 2016, saw their last elephant show, in Providence, Rhode Island. Six female Asian elephants lumbered, danced, and spun about. On cue, they stood on pedestals and on each other. After their last sashay, they linked together trunk-to-tail and disappeared behind the red curtain. Later they would ride a train to the 200-acre Ringling Bros. and Barnum and Bailey Center for Elephant Conservation in central Florida.

But the story doesn't end there. In part due to dwindling audiences, the owners of RBBX shut down the show forever in May 2017. It is the end of exploitation not only for the elephants, but also for the tigers, lions, and other trained animals that have endured whips, electric prods, and bullhooks. Watching wild animals perform tricks has become socially unacceptable.

Here Be Dragons

Picture yourself a Chinese sailor visiting the Lesser Sunda Islands of Indonesia thousands of years ago. To reach the rugged landscape you sail through churning ocean. On land, a 10-foot (3-m) lizard charges toward you. Your wild tales of narrowly escaping the slavering jaws of the giant beast help form the basis for the mythical Chinese dragon. Much later, the Hunt-Lenox Globe, built in Europe about 1503–1510, features a warning in Latin, "*hic sunt dracones* [here be dragons]," on the southeast coast of Asia.

In 1910, a Dutch soldier named van Hensbroek sailed to Komodo Island to investigate tales of these monster lizards. He shot one and sent the skin to the museum director in Java. The beast was a monitor lizard, eventually named *Varanus komodoensis*. Komodo dragons, the world's largest lizards, are found only on five Lesser Sunda Islands. On March 6, 1980, Komodo National Park was established to protect the lizards and their habitat.

The legend of Putri Naga reflects Komodo Islanders' respect for their lizards. A mythical dragon princess named Putri Naga married a man and gave birth to twins. The boy, Si Gerong, was raised among

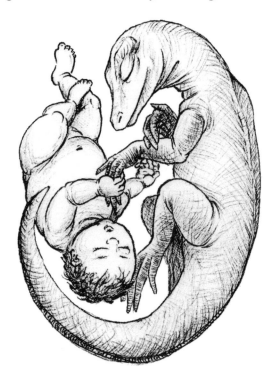

people. The other twin, a female dragon named Orah, grew up in the forest. Each was unaware of the other. Years later, Si Gerong shot a deer. A large monitor seized the deer, barred its teeth, and hissed. As Si Gerong raised his spear to kill the lizard, Putri Naga appeared and warned, "Stop! The lizard is your twin sister, Orah, your equal." The story continues that henceforth, the Komodo Islanders respected the monitors and even fed aging "brothers" unable to feed themselves. Some islanders still believe that if a dragon is harmed, its human relatives will suffer also.

MARCH 7

Farming the Hair

The 5.9-inch-long (15-cm) crustacean has tiny eyes and is probably blind. Silky blond hair-like setae cover its chelipeds (the claw-bearing legs) and claws. It lives on hydrothermal vents at a depth of 7200 feet (2200 m) in the South Pacific Ocean, south of Easter Island. It's the Yeti crab (*Kiwa hirsuta*), discovered in 2005 and announced to the world on March 7, 2006—a new species, new genus, and new family of crustacean. Kiwa is a guardian of the sea in Polynesian mythology, and *hirsuta* is the Latin word meaning "hairy." Derivation of the common name is obvious: a miniature legendary Abominable Snowman (also known as Yeti).

Since 2006, two more species of Yeti crabs have been discovered. *Kiwa puravida* lives off Costa Rica, and *Kiwa tyleri* occurs in the East Scotica Ridge in the Southern Ocean, Antarctica. The hair-like setae of Yeti crabs host chemosynthetic bacteria (bacteria that use chemical energy instead of light energy to synthesize organic compounds). Yeti crabs eat the bacteria growing on their bodies, and, amazingly, they "farm" these microorganisms. By waving their chelipeds, the crabs remove boundary layers that might otherwise limit carbon fixation, thereby increasing production of the bacteria.

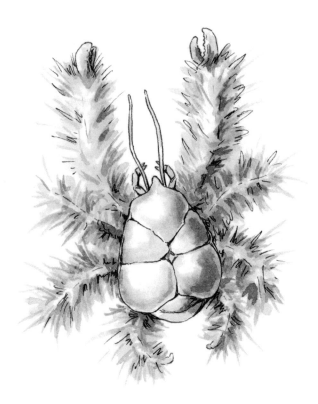

The elusive Yeti has finally been discovered! It just isn't the Abominable Snowman we imagined.

MARCH 8

Tortoise Transport

The US Marine Corps plans to expand its Air Ground Combat Center in Twentynine Palms, San Bernardino County, California. To protect the resident desert tortoises, the Corps, at an estimated cost of $50 million, intends to airlift about 1500 tortoises by helicopter from their native land to elsewhere in the Mojave Desert.

The plan sounds commendable, but biologists predict that up to 50 percent of the translocated tortoises will not survive. Previous translocation efforts for desert tortoises have shown that the stress of being handled can increase their vulnerability to disease and predators. Relocation disrupts the tortoises' social network. Some become disoriented and try to return home. As they wander, they risk being run over by cars and eaten by ravens and coyotes. On March 8, 2016, the Center for Biological Diversity filed a notice of intent to challenge the Corps' plan. Additional impact studies were carried out, and by early March 2017 the relocation plan was back on track after the US Fish and Wildlife Service determined that relocation would not jeopardize "survival of the species." Meanwhile, the Center points out that tortoises will be relocated to lands where tortoises are dying off—and we don't know why.

Sometimes our actions on behalf of wildlife, though carried out with good intentions, end up hurting the animals we are trying to help. Sometimes we don't know how best to mitigate problems we cause for wildlife. Unfortunately, sometimes our actions stem from political motives or legal obligation rather than sincere concern for the animals.

MARCH 9

Festival of Whales

Whale-watching has become a popular tourist activity worldwide, with more than 13 million people participating each year. We enjoy watching whatever behavior whales share with us: breaching, tail slapping, blowing, and playing. The more we learn about whales, the more we appreciate their intelligence, as for example reflected in some species' complex communication and teamwork to herd schools of fish.

The coast of California is a great place to watch gray whales as they migrate from their feeding grounds in the cold waters of the Chukchi and Bering Seas to warm-water lagoons of Mexico's Baja California Peninsula, where they mate and give birth. The first two weekends of March host the annual Festival of Whales in Dana Point, California, an event that has celebrated the annual migration

of gray whales for more than 45 years. The festival offers something for everyone: parade, whale-watching expeditions, clam chowder cook-off, BBQ, music, beach cleanup, marine mammal lectures, a classic car display, and kids' fishing clinics.

Today's whale conservation mantra—a whale is worth more alive than dead—has led to whale-watching opportunities from coasts around the globe: Canada to Argentina, Norway to South Africa, India, the Philippines, Japan, Australia, New Zealand, and more. The hope is that once we see these magnificent mammals frolicking in the waves, we admire them and support their conservation.

MARCH 10

"Land of Tortoises"

The Galápagos Islands, often called *Las Islas Encantadas* (The Enchanted Islands), are isolated in the Pacific Ocean about 620 miles

(1000 km) from the coast of South America. As with many of the now-resident plants and animals, which were blown in by the wind or carried in rafts of tangled vegetation, people landed on the islands by accident. In 1535, Fray Tomás de Berlanga, Bishop of Panamá, was headed to Perú to settle a land dispute between Francisco Pizarro and Diego de Almagro after the conquest of the Inca Empire. His ship drifted west from the Ecuadorian coast on the strong Humboldt Current, and he made landfall on the Galápagos archipelago on March 10, 1535.

Once the islands appeared on maps in about 1570, fortune-seeking buccaneers and pirates used the islands as a base from which to raid Spanish galleons transporting Incan riches to Spain. By the early 1800s, whalers hunted off the islands for sperm whales and other marine mammals. In 1835, Charles Darwin made his historic visit to the archipelago—exactly three centuries after Tomás de Berlanga discovered them. Darwin realized the truly enchanted nature of the islands: they are home to a vast number of endemic species and are a showcase for evolution.

By the early 1900s, the population of full-time human residents on the islands was growing. First came convicts and societal misfits, and later entrepreneurs. In 1959, the archipelago was proclaimed a national park. Now, with stringent regulations, the Galápagos Islands, "land of tortoises," welcome ecotourists to marvel at their resident Sally Lightfoot crabs, marine iguanas, land iguanas, giant tortoises, blue-footed boobies, Galápagos fur seals, and more.

MARCH 11

Rediscovery?

> This majestic and formidable species, in strength and magnitude, stands at the head of the whole class of woodpeckers hitherto discovered.
> —ALEXANDER WILSON, *American Ornithology* (1808–1814)

The sight of an ivory-billed woodpecker, at 20 inches (51 cm) long, must have been awesome. By the early 1960s, only a few sightings of this carmine-crested and polished-ivory-billed bird were reported. The decline was due mostly to extensive logging in the birds' native forests following the Civil War. On March 11, 1967, the ivory-billed woodpecker was listed as Endangered. Many thought the species was already extinct.

Then, in April 2005, birders reported seeing an ivory-billed woodpecker in a vast swamp forest in Arkansas. They videotaped a few seconds of the bird. Some ornithologists who analyzed the tape suggested the bird was the smaller pileated woodpecker. Field biologists associated with the Cornell Lab of Ornithology and its partners searched for the woodpeckers for the next five years. They covered over 500,000 acres of forest in eight states, with no sightings.

Possible rediscovery of the species, long assumed to be extinct, has captured the attention of birders worldwide. In December 2008, The Nature Conservancy announced a $50,000 reward to anyone who could lead one of its biologists to a living ivory-billed woodpecker. The woodpecker is currently classified as Critically Endangered, and the reward offer still stands. For an avid birder, there's nothing quite like the challenge of rediscovering a species thought to be extinct.

MARCH 12

Kelly

March is Dolphin Awareness Month, a time to celebrate these highly social and intelligent animals. Kelly the Dolphin from the Institute for Marine Mammal Studies in Gulfport, Mississippi, is teaching us another angle to dolphin intelligence. All the dolphins housed at the Institute are trained to gather and hold litter that falls into their pools. When they see a trainer, they surrender the litter and get

rewarded with fish. Kelly, however, has changed the game rules. When she finds paper in her pool, she hides it under a rock. Once she spots a trainer, she dives down, tears off a piece, gives it to the trainer, and gets her fish reward. Then she dives back down, rips off another piece, gives it to the trainer, and scores another fish. She continues until the last scrap has been surrendered. Kelly has a sense of the future and has learned how to manipulate the system.

That's not all. One day a gull flew into her pool. Kelly grabbed it, waited for the trainer, and handed it over. The gull was large, so the trainer gave Kelly lots of fish. The next time Kelly was fed, she hid the last fish under the rock where she hides paper. When the coast was clear, Kelly used the fish as bait to lure a gull. The trainer rewarded her with lots more fish. She taught the trick to her calf, who taught it to other calves. Gull-baiting is now a popular game among the Gulfport dolphins.

MARCH 13

Our Mysterious, Precious Kidneys

> All quadrupeds that are viviparous have kidneys, but of those
> which are oviparous the tortoise is the only one that has them.
> —PLINY THE ELDER, *Naturalis Historia* (77–79 CE)

One wonders where Pliny got his misinformation, but he is not alone in misunderstanding kidneys, long viewed as mysterious. In ancient Egypt, the only organs left inside mummified bodies were the heart and kidneys. Some scholars suggest that the Egyptians associated kidneys with judgment and thus kept them intact for the deceased's afterlife. The word "reins" is often used metaphorically in Judeo-Christian writing in place of kidneys. Reins were regarded as the seat of conscience, desires, and affections. The Talmud teaches that one kidney counsels what is good, the other what is evil. The Bible teaches that God evaluates a person's goodness by inspecting his kidneys and heart: "For the righteous God trieth the hearts and reins" (Psalms 7:9).

The second Thursday in March is World Kidney Day, begun in 2006 to increase awareness of the importance of kidneys to our overall heath and the need to reduce kidney disease, and to educate people of the valuable lifesaving gift they can leave behind by becoming an organ donor. You could celebrate World Kidney Day by indulging in cherries, cranberries, olive oil, onions, and garlic, all purported to promote kidney health. And maybe sign up as an organ donor if you haven't already done so?

MARCH 14

The Spider's Web of Life

March 14 is Save a Spider Day, a time to appreciate spiders for their service of eating flies, mosquitoes, and cockroaches in our homes, and aphids, caterpillars, and grasshoppers in our gardens. Virtually all cultures tell sacred stories that explain the creation of Earth and all of life. Perhaps the highest honor a culture can bestow upon an animal is the creator role. The spider holds that mythological position in many parts of the world, for it is a natural creator. Working somewhat like human fingers, spinnerets on the tip of the abdomen extrude silk from glands. Many spiders then weave the silk into intricate—and dangerous—webs.

The Hopi of the southwestern United States feature the spider in their creation story. In the beginning there was only Spider Woman and Tawa. Spider Woman controlled the earth; Tawa controlled the heavens. After Spider Woman made plants and nonhuman animals, she formed people by mixing yellow, red, white, and black earth with her saliva. She covered them with the Cape of Wisdom, cradled them, and chanted the Song of Life. Once the people breathed life, she taught them traditional ways of living. She instructed men to hunt, grow corn, and make offerings to the gods. Women should watch over their homes and lead the families. She also taught the people the art of weaving. Spider Woman wove together all living beings with her silk, creating our interconnected web of life.

MARCH 15

FLAP

Thud. Another cedar waxwing has crashed into your window and is now lying prostrate on your patio. Window strikes are the leading cause of bird mortality across North America, with an estimated 100 million to nearly 1 billion birds dying each year. A bird sees trees, sky, or clouds mirrored in the glass and flies directly into the window. If it doesn't die from internal injuries, a stunned bird on the ground is easy prey for neighborhood cats.

We love to feed the birds, but our feeders cause bird/window collisions. Stand next to your bird feeders and get a birds'-eye view of your windows. If you can see trees, sky, or clouds reflected in the glass, birds can too. Try moving the feeders elsewhere. Or break up the windows' reflections with visual cues or markers that alert birds to the glass.

The first organization to address bird/building collisions is FLAP (Fatal Light Awareness Program) CANADA, founded in 1993. Check out their website, www.flap.org, for ideas on how to make your windows visible to local and migrating birds. When you click on the site you'll see a box with advancing numbers that indicate the estimated number of migratory birds that have died in window collisions across North America since you opened the page. On March 15, 2016, during the first 60 seconds I spent reading about FLAP, an estimated 2375 birds across North America died from crashing into windows.

MARCH 16

Icon for Conservation

> Shrouded in bamboo was a giant panda, a female, slumped softly in the snow, her back propped against a shrub. Leaning to one side, she reached out and hooked a bamboo stem with the ivory claws of a forepaw, bent in the stem, and with a fluid movement bit it off near the base. Stem firmly grasped, she sniffed it to verify that it was indeed palatable, and then ate it end-first like a stalk of celery.
> —GEORGE SCHALLER, *The Last Panda*

Today is Panda Day, a celebration of these black and white bears that have captured the hearts of people everywhere. Pandas remind us of furry versions of ourselves, as they eat while sitting on the ground, manipulating bamboo with impressive dexterity. Their black eye patches make their eyes seem large and remind us of baby animals, including our own. Although giant pandas may once have numbered 100,000 animals, the most recent census (2014) revealed only 1864 individuals remaining in the wild. Now they live in mountainous areas, having been driven out of the lowlands because of deforestation and habitat fragmentation.

Outside of China, giant pandas were virtually unknown until the 1800s. The Western world first heard about giant pandas in 1869 when French missionary-naturalist Père Armand David described a panda from a specimen shot by hunters in Sichuan Province. Times have changed. Now toddlers worldwide cuddle plush pandas, and children dress as pandas for Halloween. Because pandas are so well loved, they have become an icon for the conservation of all plants, animals, and ecosystems.

MARCH 17

Symbolic Snakes

One day in the fifth century CE, Irish raiders took a 16-year-old boy named Patrick from his home in Roman Britain to Ireland, where they enslaved him as a shepherd. After six years, Patrick escaped, returned to Britain, and later studied religion in France. Eventually he returned to Ireland, where he vowed to convert the pagan Irish to Christianity. According to legend, Patrick was absorbed in a 40-day fast atop an Irish mountain when snakes attacked him. Patrick drove the snakes into the sea and rid Ireland of snakes forever. (Of course Ireland never had any snakes, due to its geographical location and glacial history.)

The legend of St. Patrick is an allegory, in which snakes represent early pagan faiths brought to the island by invading peoples and traders. Snake worship was one of the most offensive pagan practices to the early Christians, because the Church demonized snakes as personifications of the Devil: It was the evil serpent who tempted Eve to eat the forbidden fruit in the Garden of Eden and

caused the "fall of man." In contrast, for people of various Celtic religions, the snake symbolized fertility, knowledge, healing, rebirth, and immortality—all that is good. Although he did not literally drive snakes out of Ireland, Patrick made sizable inroads to squelching Irish pagan religious practices—the symbolic snakes.

March 17 is St. Patrick's Day, a religious and cultural celebration held on Patrick's traditional death date. The day is a public holiday in both Northern Ireland and the Republic of Ireland, a time to celebrate the ousting of the symbolic snakes.

MARCH 18

Silk Threads

According to an ancient Chinese legend, one day the Empress Lei-Zu was drinking tea under a tree when a cocoon fell from a branch into her cup. She removed the object and when the cocoon began to unwind from the hot water, she wrapped the silk thread around her finger. Once the thread ran out, she found a little caterpillar on her hand and realized the larva had made the silk. She told her story to the people, and that was how the Chinese learned to harvest silk.

The Chinese have been breeding mulberry silkworms (domesticated silk moth caterpillars) to produce silk for at least 5000 years. Chinese emperors guarded knowledge of silk production to maintain a monopoly, but word leaked and reached Korea by around 200 BCE. Later the practice spread to Japan, India, and eventually the West. Silkworm larvae eat continuously, molt four times, and then enclose themselves in a cocoon spun from a single thread of raw silk up to 3000 feet (900 m) long. The pupae release proteolytic enzymes to make exit holes, from which they emerge as adult moths. These enzymes destroy the silk. Thus, to preserve the silk for harvest, workers boil the cocoons and kill the pupae.

The Silkworm Festival near Phnom Penh, Cambodia, is celebrated on the day of the full moon in March. A fashion show features gorgeous silks made by local women, men, and children. More importantly, the festival gives thanks to the silkworms sacrificed over the previous year.

MARCH 19

Return to Capistrano

The Legend of Capistrano tells that cliff swallows took refuge in the city's Spanish mission after an innkeeper destroyed their mud nests. Each year, the swallows return to San Juan Capistrano, California, on March 19 (St. Joseph's Day) to rebuild their nests and raise their chicks in the protection of the church ruins. The legend continues that each October 23 (Day of San Juan) they bid farewell and head south again.

Cliff swallows do in fact return to southern California in late February or March. They fly the 6000 miles (10,000 km) north from Argentina, where they have vacationed for seven months during the austral summer. Year after year, flocks of swallows descended on the mission. Since 2009, however, the flocks have not appeared. Instead, the birds have found good nesting sites north of San Juan

Capistrano. Local urbanization and repairs done to the church may have caused the shift. In 2016, artificial mud nests were attached to the mission walls in hopes of luring the birds back. Time will tell if this game plan works.

Whether the swallows return or not, every year the village of San Juan Capistrano holds a week-long *Fiesta de las Golondrinas* (Swallows Festival). Beneath the gaiety of BBQ, parades, and street fairs, participants' hearts and minds focus on the little migratory bird, the cliff swallow, symbol of faithfulness and devotion to home.

MARCH 20

The Many Ways to Beget a Frog

March 20 is World Frog Day. There are many reasons to appreciate frogs, including their propensity to eat mosquitoes and grasshoppers, but as an amphibian behavioral ecologist, I take special pleasure in bragging about the many ways frogs reproduce.

Glass frogs lay their eggs on leaves above water; the hatchling tadpoles fall into the water and continue to develop. Poison dart frogs deposit eggs on land; after hatching, the tadpoles climb onto one of the parents and ride piggyback to water. Male midwife toads wrap the strings of fertilized eggs around their hind legs. When time comes for the eggs to hatch, the males dip into a pond or puddle. Male Darwin's frogs brood their tadpoles in their vocal sacs, and female marsupial frogs brood their young in dorsal pouches. Female gastric-brooding frogs (2 species, both believed to have gone extinct in the 1980s) swallowed their eggs; the young developed and metamorphosed in the mothers' stomachs. Dink frogs lay their eggs on land and skip the tadpole stage; miniature froglets emerge from the eggs. In some dink frogs, a parent guards the eggs and moistens them with bladder water. Females of at least nine species of frogs retain the young in their oviducts until they are born as miniature froglets.

As part of their transition from water to land, amphibians evolved the greatest reproductive diversity of all tetrapod vertebrates. That's something to croak about on World Frog Day!

MARCH 21

Fancy Chickens Replace Rattlesnakes

Daredevils kiss rattlesnakes and crawl into rattlesnake-occupied sleeping bags. Spectators gawk into enclosures writhing with rattlers. Vendors sell rattlesnake meat and skins, and key chains made of rattles. These scenes are all part of a rattlesnake roundup.

In preparation for the festivities, collectors scour the landscape and capture every rattler found, most of which will be slaughtered. Rattlesnake roundups are superficially legitimized by civic or charitable organizations as fundraisers, touted as a public service for "controlling" rattlesnake populations and reducing the incidence of snakebite. In reality, fewer rattlesnakes mean more rodents. The events persecute the snakes and exploit the public's fear of venomous snakes. Rattlesnake roundups date back to 1939 and have been held in over a dozen US states. A backlash against the cruelty of rattlesnake roundups has encouraged some communities to develop other ways to attract tourist dollars.

Fitzgerald, Georgia, has invented a unique replacement for their rattlesnake roundup—the Fitzgerald Wild Chicken Festival, held on the third weekend of March. The festival celebrates the town's unique Burmese chickens. In the 1960s, the Georgia Department of Natural Resources introduced nonnative Burmese chickens across the state as a flashy game species. The only place the fancy chickens thrived was Fitzgerald, where they roam backyards and gardens. The Fitzgerald Wild Chicken Festival attracts about 10,000 people each year, far more than the rattlesnake roundup ever did.

Shape Shifters

The mournful cries and soulful expressions of seals render them almost human. These qualities likely gave rise to the Irish, Scottish, and Icelandic myths of "selkies"—enchanted beings that are seals by day and, after shedding their skins, humans by night. A selkie-man in human form is said to possess magical, seductive powers over women. A selkie-woman is irresistible to men.

March 22 is International Day of the Seal. In addition to celebrating seals, why not honor the selkies of folklore fame? Prepare yourself for tears while reading selkie tales; most are romantic tragedies. Once the selkie returns to the sea, he or she seldom, if ever, visits land or the human partner again, as in the following story from Orkney, Scotland.

One evening a young man watched a group of seals come ashore.

They cast off their skins and became alluring young women. The man picked up a sealskin, and the young woman to whom the skin belonged followed him home. She was now his captive. The woman soon forgot her life in the sea, content in her human form. The couple married and in time had three children. Many years later, the youngest child discovered a strange bundle in the house and took it to his mother. She recognized it as her sealskin and remembered her life in the sea. She climbed back into her skin, said goodbye to her human family, returned to the sea, and never again came ashore.

MARCH 23

Wildness

World-renowned nature photographer Eliot Porter (1901–1990) began photographing with a Brownie box camera in 1912. In his words, he "found the tonic of wildness" on his family's island in Penobscot Bay, Maine—Great Spruce Head Island. It was also there he first read Thoreau's *Walden*. Ten years out of medical school, Porter traded medicine and science for photography. His wife, Aline, a painter, suggested he combine his photographs with Thoreau's writings because she found the two so complementary. Porter pursued her suggestion, and off and on over ten years he photographed the woods, streams, ponds, and marshes of the northeastern United States, in all seasons.

In 1962, the Sierra Club published Porter's masterpiece *In Wildness Is the Preservation of the World*. The book pairs Porter's color photographs from New England with Thoreau's entries in *Walden*. The entry for March 23, 1856, "I seek acquaintance with Nature,— to know her moods and manners," sums up what Thoreau wished to do and what Porter's photography aimed to reflect. Porter's accompanying photograph captures crimson trillium reaching out from a bed of dry leaves. In the introduction to Porter's book, environmentalist David Brower noted that Porter and Thoreau traveled closely together, a century apart. Porter's book was published 100 years after Thoreau's death in May 1862. Porter's photography reflects and endorses what Thoreau discovered so long ago—that wildness is necessary for the health of the human spirit.

MARCH 24

Nature in the New World

Imagine being an early eighteenth-century European naturalist, wondering what forms of life might be found in that "New World" across the ocean. Someone was destined to explore this exotic landscape and share knowledge of this New World with the Old World. That person was Mark Catesby, born in England on March 24, 1682 or 1683.

A self-taught artist and naturalist, Catesby visited eastern North America from 1712 to 1719 and again from 1722 to 1726. Catesby explored widely, took copious notes, and painted from life whenever possible to illustrate the plants and animals of this New World. After returning to England in 1726, he spent the next 20 years working on his two-volume *Natural History of Carolina, Florida, and the Bahama Islands,* the first published account of the flora and fauna of North America. Catesby was the first to use folio-sized colored plates in a natural history book, but it didn't come easily. To print the watercolors he made in America, the illustrations had to be engraved onto copper plates. Lacking the resources to pay experts to accomplish this task, Catesby learned the technique and executed the 220 hand-colored etchings himself.

We owe much to Catesby, called by some the greatest British natural history explorer of his time. His vivid descriptions and stunning artistic renderings of nature in eastern North America encouraged further scientific exploration. Catesby is immortalized in many scientific names of North American plants and animals, including the American bullfrog, *Lithobates catesbeianus.*

MARCH 25

Festival of Color

> Colors are the smiles of nature.
> —LEIGH HUNT, nineteenth-century English poet

How we love color! Holi, the ancient Hindu festival of color, celebrated in India, Nepal, and elsewhere in southern Asia, is a time to welcome spring with color, friendship, music, and dance. The festival date, determined by the Hindu calendar, typically falls in March. Festival-goers dress in bright colors, throw colored water on each other, and smear each other with colored powders.

Mother Nature must have been smiling when she clothed some animals in outrageous colors. But she had her reasons. The mandrill's face, with purple and blue ridges framing his bright red nose, matched by his red and blue rear end, and the peacock's showy emerald tail feathers sporting "eye" markings of blue, gold, and red are meant to impress the ladies. Chameleons and cuttlefish shuttle back and forth from attention-grabbing blues, greens, reds, and yellows to communicate with each other, to subdued grays, browns, tans, and blacks for camouflage. Distasteful or toxic marine nudibranchs warn potential predators with bright colors: blue, green, yellow, and black; pink and purple; purple and yellow. Poison frogs do the same with their granular skins of blue, red, orange, turquoise, green, and gold, often juxtaposed with black. Humans are rather drab animals, but at least we have the option of wrapping ourselves in color.

MARCH 26

A Catalog of Animals

Conrad Gesner, physician and foremost naturalist of the sixteenth century, was born in Zürich, Switzerland, on March 26, 1516. His magnum opus, the five-volume *Historiae animalium* (published 1551–1558 and 1587), is considered the beginning of "modern" zoology. Gesner joined a few others at the time in reviving the practice of direct observation of nature. Up to that point, scholars had relied heavily for their writings on the works of classical intellectuals such as Aristotle, Pliny the Elder, and Aelian.

Gesner meant for *Historiae animalium*, a 4500-page encyclopedia of animals, to provide an accurate and comprehensive catalog of the animal kingdom. He sought to distinguish real animals and observed facts from mythological animals like unicorns, mermaids, and basilisks. His volumes were illustrated with more than 1000 hand-colored woodcuts made from observations by himself and his colleagues, and from other sources including Albrecht Dürer's famous "Rhinoceros." The volumes were the first categorical attempt to illustrate animals in their natural environments. Many of the drawings are so detailed that the animals can be identified to species.

Story has it that when Gesner knew his life was near the end, he asked to be taken to his beloved library, to die among his books. And there he died of the plague in 1565, at age 49, a year after he was inducted into the noble class.

MARCH 27

Sacred Domain

In the past, the lives of many Tlingit of coastal southeastern Alaska revolved around harvest from the sea—fish, marine mammals, shellfish, and seaweed. By late March, they began fishing for halibut, a rite of passage into a sacred domain, venturing miles from shore into the open sea where they risked sudden ocean squalls. Halibut fishing was dangerous also because the fish are large and powerful, growing to 120 pounds (54 kg) or more.

To assure safe passage and a successful catch, Tlingit fisher-

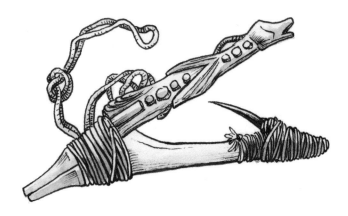

men relied on supernatural assistance in the form of halibut hooks carved with symbolic images. They lashed together two pieces of cedar into a V about 12 inches (30 cm) long. One arm contained an iron barb. The other was carved with an image of a powerful animal that embodied a spirit helper, often a raven, river otter, diving bird, octopus, halibut, or human, which transformed the hook into a potent charm. The spirits of the carved images were believed to entice halibut to the bait. Fishermen entreated their hooks to catch halibut and, once caught, to hold tight.

Today, Tlingit fishermen still respect traditional practices not only when fishing, but also when preparing the fish. The Tlingit view halibut as special, the first fresh fish of the year, a welcome change from the dried fish they eat all winter.

MARCH 28

Sacred Cats

> What greater gift than the love of a cat?
> —CHARLES DICKENS

> There are two means of refuge from the miseries of life—music and cats.
> —ALBERT SCHWEITZER

Around 2600 BCE, Egyptians domesticated the wildcats that were attracted to their towns by the presence of mice and rats. People likely left food to encourage the cats to stick around and protect their granaries from the rodents. Eventually cats moved inside, where they found warmth and a reliable source of food.

Egyptians worshipped cats as incarnations of Bast (Bastet), goddess of fertility, protection, and motherhood. Bast typically was depicted as a woman with a cat's head, and she carried a four-stringed rattle to ward off evil spirits. Egyptians considered cats sacred and treated them with ultimate respect. Cats were housed in Bast's temples, and mummified cats were left as offerings in hopes of securing fertility and protection.

Although today most of us don't literally worship our cats, many people find them to be superb companions. The Humane Society of the United States estimates that there are 86.4 million cats kept as pets in the US. Nearly 35 percent of homes have at least one cat. March 28 is Respect Your Cat Day—the perfect day to give your feline some extra special attention.

MARCH 29

Dancing with Daffodils

> I wandered lonely as a cloud
> That floats on high o'er vales and hills,
> When all at once I saw a crowd,
> A host of golden daffodils;
> Beside the lake, beneath the trees,
> Fluttering and dancing in the breeze.
> —WILLIAM WORDSWORTH, "I Wandered Lonely as a Cloud"

If you were born in March, the lovely golden daffodil is your birth flower. Daffodils (genus *Narcissus*) announce the arrival of spring with their trumpet-like petals. We associate daffodils with rebirth

and new beginnings. They symbolize happiness, respect, and friendship. As a symbol of life itself, the daffodil represents hope that a cure for cancer will be found. Every March, cancer societies around the world, including the Canadian Cancer Society and the American Cancer Society, feature daffodils in fund-raising events to support the cure for cancer.

In ancient Greek and Roman mythology, daffodils represented unrequited love. One myth tells that a nymph named Echo saw 16-year-old Narcissus, son of the river god Cephissus, walking in the woods. Narcissus was renowned for his beauty, and Echo fell madly in love. He spurned her, leaving her heartbroken until only an echo of her voice remained. Narcissus drank from a pool and saw his reflection for the first time. Amazed at his beauty, he fell in love with himself. Unable to have the object of his desire, he wasted away. After his body was gone, a yellow narcissus (also known as daffodil) grew in its place.

MARCH 30

Bunny Footprints

When I was a child in Pennsylvania, my neighbor arose before daybreak Easter morning and tiptoed outside with a cup of flour. Using her fingertips, she made little rabbit footprints running up and back down the driveway. The Easter bunny had come and gone.

The seemingly absurd concept of an Easter bunny that brings

children chocolate eggs and decorated hard-boiled chicken eggs likely arrived on American shores with Protestant German immigrants who settled in Pennsylvania in the 1700s. In Germany, children made nests in their caps and bonnets before Easter. If they had been well behaved, the egg-laying hare called *Osterhase* left colored eggs in the nests. In the US, the wild Easter hare got supplanted by the more docile Easter bunny, and in time the Easter bunny brought chocolates and other candy in addition to colored eggs. Rabbits and hares, with their well-known reputation for prolific breeding, have long symbolized fertility, and eggs have long represented new life. Springtime and Easter are a time of rejuvenation, new life being born, new beginnings, and renewed energy. Symbolically, then, the rabbit-and-egg combination seems natural for Easter, the Christian celebration of Christ's resurrection.

Easter is celebrated the first Sunday after the full moon following the spring equinox. The calculation varies between eastern and western parts of the world, which explains why the Easter bunny makes his appearance in Germany on a different day than he visits children in Pennsylvania.

MARCH 31

The Music of Nature

Just as nature inspires literature and the visual arts, it permeates music. Throughout history, classical composers have incorporated the sounds of nature—and the pitches, rhythms, and textures of those sounds—into some of our best-loved classical music. Take for example "The Frog" quartet, written by Joseph Hayden. The string quartet No. 41 in D Major, Opus 50 No. 6 is often called "The Frog" because of the croaking frog in the finale, created by the first violinist playing the same notes alternately on two neighboring strings. March 31 is the birthday of Joseph Hayden, born in Austria in 1732.

Other images of nature in classical music include a barking dog, summer thunderstorm, rustling leaves, and falling snow in Vivaldi's "Four Seasons"; reflection of the moon on Lake Luzerne in Beethoven's "Moonlight Sonata"; bird trills in Vivaldi's flute concerto "Il Gardellino," nicknamed "The Goldfinch"; and ocean waves in Debussy's "La Mer." Rossini's "William Tell Overture" features galloping horses, and Rimsky-Korsakov's "Flight of the Bumblebee" mirrors the frantic, chaotic flying of a bumblebee. Biber's "Sonata Representativa" projects the sounds of a cuckoo, nightingale, hen, rooster, quail, frog, and cat. Flowing water, bird songs, and a building thunderstorm accompanied by howling gales and lightning are central to Beethoven's Symphony No. 6, also known as the Pastoral Symphony. And, of course, diverse animal sounds are represented in Prokofiev's "Peter and the Wolf" and Saint-Saens' "Carnival of the Animals." Human creativity naturally reflects our surroundings.

APRIL 1

April Fool!

For some animals, once detected by a predator, the most effective defense is surprise or trickery. Opossums and hognose snakes play dead. Horned lizards and rabbits hunker motionless. Praying mantids do the opposite: they stand tall, fan out their wings, and spread their front legs in a threatening pose, appearing larger than life. Skunks spray foul odors, fulmar chicks regurgitate, toads pee, garter snakes defecate, llamas spit foamy saliva, and electric eels electrocute. Crabs, starfish, and lizards break off legs, arms, or tails, respectively, leaving the wiggling part behind as they escape.

Some animals fool predators by mimicking unpalatable objects or large or dangerous animals. Wood nymph caterpillars resemble piles of fresh bird poop. The colors and shapes of katydids resemble twigs or munched-on leaves. Some octopuses resemble rocks; others change shape and color, seemingly becoming venomous sea snakes. Nymphs of some West African assassin bugs pile the corpses of their prey—ants, termites, and other insects—onto their backs. If a predator sees through the deception, the nymph ditches its cadaverous backpack and flees. Owl butterflies, luna moths, and four-eyed butterfly fish flash large, false "eyespots," giving the impression of a much larger animal.

Sometimes nature follows Niccolò Machiavelli's advice: "Never attempt to win by force what can be won by deception" (*The Prince*, 1532). Happy April Fool's Day!

Once Upon a Time

> Reading one book is like eating one potato chip.
> —DIANE DUANE, *So You Want to Be a Wizard*

What was your favorite book as a small child? *The Story of Ferdinand? The Very Hungry Caterpillar? If You Give a Mouse a Cookie? The Cat in the Hat? Winnie the Pooh? Curious George?* How about when you were a little older? *Doctor Doolittle? Stuart Little? The Lion, the Witch and the Wardrobe? Charlotte's Web? Olivia? The Jungle Book? The Call of the Wild? Hans Christian Andersen's Fairy Tales and Stories?* Worldwide, many favorite children's books feature animals, and it's no wonder. Children form lasting friendships with storybook animals.

International Children's Book Day has been observed every year since 1967 on or around Hans Christian Andersen's birthday, April 2, with the dual purpose of inspiring a love of reading and highlighting children's books. The sponsor is the International Board on Books for Young People (IBBY), a nonprofit organization of book-lovers worldwide committed to bringing books and children together. IBBY, founded in Zurich, Switzerland, in 1953, now has 75 national sections worldwide. Their motto is "Every child has a right to become a reader." The day is celebrated in various ways, from events with authors of children's books to writing competitions, and of course involves reading to kids.

Take a Hike

> At present, in this vicinity, the best part of the land is not private property; the landscape is not owned, and the walker enjoys comparative freedom. But possibly the day will come

> when it will be partitioned off into so-called pleasure grounds,
> in which a few will take a narrow and exclusive pleasure only,
> —when fences shall be multiplied, and man traps and other
> engines invented to confine men to the *public* road; and
> walking over the surface of God's earth, shall be construed
> to mean trespassing on some gentleman's grounds.
> —HENRY DAVID THOREAU, "Walking" (1862)

Henry David Thoreau walked at least four hours each day. He found that if he didn't walk, he couldn't write. Unfortunately, Thoreau's premonition of public land being gobbled up by private landowners has come to pass. Today, over 60 percent of US land is privately held. In some states, that figure is over 90 percent. No Trespassing and Private Property signs warn us to keep out.

In contrast, in some European countries it is perfectly acceptable to walk through private land. The Scottish call it "the right to roam." The 2003 Scottish Land Reform Act opened up the entire country for mountain biking, horseback riding, canoeing, swimming, and camping, as long as activities are carried out "responsibly."

The first Wednesday in April is National Walking Day, sponsored by the American Heart Association—a day to inhale some fresh air and enjoy your surroundings. But beware of gun-toting landowners should you trespass. We don't have a right to walk through much of the United States of America.

APRIL 4

Rats Rule

Ratlist, an online discussion site where people ask and answer questions about their pet rats, began in 1999. As of August 2017, more than 2300 rat-owners belong to Ratlist. In 2002, members proposed World Rat Day to celebrate rats as intelligent and affectionate pets. World Rat Day is still celebrated each April 4.

Rats rule every day at the Karni Mata Temple in Rajastan, India, an ornate Hindu temple in which about 20,000 black rats (*Rattus rattus*) are fed and revered. According to legend, a Hindu warrior deity of the Charan caste of storytellers, named Karni Mata, tried to revive the dead son of one of her storytellers. In return for the boy's life, Karni Mata promised the god of death that all male storytellers would be reincarnated as rats in her temple. When each temple rat dies, it is reincarnated back into the Charan caste. Hindus from throughout India visit the temple to pay their respects. Shoes are forbidden inside the temple, where guests hope a rat will run across their feet for good luck. The ultimate hope is seeing one of the few resident white rats, considered to be manifestations of Karni Mata and her kin. Visitors are careful not to accidentally step on and kill a rat, for if one is killed, the perpetrator must purchase a silver or gold rat figurine to be left in the temple as atonement.

APRIL 5

Cherry Blossom Time

Since 1935, Washington, DC, has hosted an annual Cherry Blossom Festival to honor the friendship between the United States and Japan. Peak flowering time varies, but is often on or around April 5. Visitors come from all over the world to admire the 3800 trees clothed in delicate white and pink flowers.

The trees came from Japan, where cherry blossoms (*sakura*) represent the beauty and frailty of life—wondrous, but tragically short. In April 1909, Dr. Jokichi Takamine (the Japanese chemist who discovered adrenaline) was visiting Washington. When he heard that cherry trees from a Pennsylvania nursery would soon be planted along the Potomac River, he asked Helen Taft, the President's wife, if she would accept a donation of an additional 2000 cherry trees, in friendship, from the city of Tokyo. Mrs. Taft accepted the offer, and in early January 1910, 2000 cherry trees arrived in Washington from Japan. After the Department of Agriculture discovered that the trees were infected with insects and nematodes, the trees were burned to protect American growers. The mayor of Tokyo offered another donation—this time 3020 trees. Healthy trees arrived in March 1912. (The original trees from the Pennsylvania nursery were a cultivar and have long since died.)

Cuttings used in propagation preserve the genetic lineage of the Japanese trees. In 2011, about 120 propagates from trees sent in 1912 were sent to Japan, continuing the cycle of giving that fulfills the trees' role as a symbol of friendship.

APRIL 6

Thank a Bat

Throughout the summer, just before sunset over 1.5 million Mexican free-tailed bats explode out from under the Congress Avenue

Bridge in Austin, Texas. The cloud flies into the night to forage on moths, gnats, mosquitoes, and other flying insects, creating an otherworldly sight of dark, flapping bodies against a blue-gray sky.

The first full week in April is Bat Appreciation Week in the United States. Bats eat crop-damaging insects and reduce the amount of pesticides we need to use. They eat mosquitoes that carry diseases like West Nile Virus and Zika. Bats pollinate bananas, guavas, mangos, and agave, and they disperse seeds and aid forest regeneration.

We appreciate bats for other reasons as well. For the Chinese, five bats grouped together represent five blessings: peace, a virtuous life, natural death, longevity, and riches. From ancient times through the present, bats featured on Chinese tapestries, art objects, and house furnishings symbolize happiness and good fortune. Fruit bats ("flying foxes") are considered sacred in the Kingdom of Tonga and are the official property of the King. Finnish folklore tells that bats are the souls of sleeping people, allowing the human spirit to explore the nighttime; the approach of a bat is welcomed as the soul of a friend or family member. In his autobiography (volume 1), Mark Twain wrote: "A bat is beautifully soft and silky; I do not know any creature that is pleasanter to the touch or is more grateful for caressings, if offered in the right spirit."

APRIL 7

Reconnect with Nature

> The world is too much with us; late and soon,
> Getting and spending; we lay waste our powers;
> Little we see in Nature that is ours;
> We have given our hearts away, a sordid boon!
> —WILLIAM WORDSWORTH, "The World Is Too Much
> with Us"

William Wordsworth, one of the best-loved English Romantic poets, was born on this day in 1770. In 1798, he and Samuel Taylor Coleridge together published *Lyrical Ballads*, a collection of poems that marked the beginning of the Romantic Movement in English literature. The poets' aim was to bring poetry to everyone, by using everyday language. A major theme of many poems in *Lyrical Ballads* was living a simpler life by reconnecting with nature.

Wordsworth continued this theme in much of his later poetry. In "The World Is Too Much with Us," published in 1807, Wordsworth chides his fellow citizens living during the Industrial Revolution for materialism, for being too absorbed in "getting and spending." He implies that people distance themselves from nature because they can't own her. He encourages people to bond with nature and enjoy her beauty—the sea, the wind, the flowers. Wordsworth's philosophy expressed in this poem is even more relevant for us today, when it is so easy to drown in consumerism, with online stores poised to sell us anything and everything, 24 hours a day.

APRIL 8

Draw a Bird

An Australian Aboriginal myth tells that a long time ago, a rainbow arching over Earth shattered into a thousand pieces. As the fragments fell, each a different color, size, and shape, they transformed into birds of red, orange, yellow, green, blue, indigo, and violet. The birds sang joyously as they tested their wings. Kookaburra broke into peals of laughter, and Eagle-hawk tried to reach the sun. April 8 is Draw a Bird Day. If you need inspiration, perhaps this myth will spark your creativity.

In 1943, seven-year-old Dorie Cooper visited her uncle in an English hospital. Her uncle had lost his leg to a land mine. To cheer him up, Dorie asked him to draw a bird. She laughed at his drawing and said that even though it wasn't very good, she would hang it in her bedroom. Dorie's honesty and acceptance lifted her uncle's spirits and those of other wounded soldiers. Every time Dorie visited, the soldiers drew bird pictures. Before long the walls were plastered with colorful birds. Three years later, Dorie was hit and killed by a car. Bird images drawn by soldiers, nurses, and doctors from her uncle's hospital ward were placed in her coffin.

Since Dorie's death, people around the world have been drawing birds on this day, her birthday, to celebrate birds and to honor a little girl who brought hope to wounded soldiers. The idea is simple: draw a bird and share it with someone.

APRIL 9

Worm Gruntin'

Come to historic downtown Sopchoppy, Florida, for the annual Sopchoppy Worm Gruntin' Festival, held the second Saturday in April. Worm gruntin' (also known as worm charming) is a time-honored technique for collecting earthworms for fishing bait. More recently, it has become a competitive sport.

How to grunt? Drive a wooden stake into the ground. Then rub the stake with a piece of iron, or drag a dulled saw across the top. Resulting vibrations in the soil make worms flee their burrows and head for the surface. The vibrations likely resemble those made by moles, which prey on earthworms, and thus elicit the worms' escape response. Humans aren't the only animals that grunt for earthworms. Many species of birds, such as seagulls, stomp the ground by alternately raising and lowering each foot. Wood turtles also stomp for worms by pounding the ground, alternating between their two front feet.

The Sopchoppy Festival has been held since 2000. In addition to worm gruntin', the festival offers craft and food vendors, live music, a ceremony to crown the annual Worm Gruntin' King and Queen, and the Worm Grunters' Ball. If Canada is more convenient, the Great Canadian Worm Charming Championship is held in Ontario in June. For Europeans, the village of Blackawton in southwest England hosts the International Festival of Worm Charming (accompanied by the Real Ale Beer Festival) in early May, and Cheshire, England, hosts the World Worm Charming Championships in June.

Plant a Tree

> Trees are the earth's endless effort to speak to the listening heaven.
>
> . . .
>
> The tree is a winged spirit released from the bondage of seed, pursuing its adventures of life across the unknown.
>
> —RABINDRANATH TAGORE, Bengali poet and philosopher; *Fireflies*

Trees provide oxygen and absorb carbon dioxide. They reduce run-off and erosion, break the wind, and reduce noise pollution. Trees provide homes for animals. Fruit trees offer food for wildlife and for us. Trees inspire us with their grace and beauty, strength, and longevity.

In 1854, journalist Julius Sterling Morton and his new bride moved from Michigan to the Nebraska territory. Both loved nature, and they soon planted flowers, shrubs, and trees on their 160-acre property. Morton became editor of Nebraska's finest newspaper, the *Nebraska City News*. He regularly wrote articles to encourage others to plant trees as windbreaks and to prevent soil loss on the nearly treeless plains. In time Morton established the first American Arbor Day, in Nebraska on April 10, 1872—a day to plant trees and appreciate their value. On that day, an estimated 1 million trees were planted in the state. Eventually, residents of all 50 US states celebrated Arbor Day by planting trees, the date depending on climate.

In 1970, President Nixon proclaimed the last Friday in April as National Arbor Day. Similar tree-celebration days are held worldwide, from Australia to Zambia. Variations include "Tree-Loving Week" in Korea, "Greening Week" in Japan, National Tree Planting Day in Tanzania and Uganda, and Dia del Árbol in Costa Rica and Mexico.

Pleasure and Farewell

> Here are sweet peas, on tip-toe for a flight:
> With wings of gentle flush o'er delicate white.
> And taper fingers catching at all things,
> To bind them all about with tiny rings.
> —JOHN KEATS, "I Stood Tip-toe upon a Little Hill"

The English Romantic poet John Keats (1795–1821) is thought to have been the first to use the name "sweet pea" for *Lathyrus odoratus*, a climbing garden plant imbued with perfume. Sweet peas have been cultivated since the seventeenth century. Wild plants have deep purple flowers; cultivars range from soft pastel blue, pink, purple, and white to two-toned beauties. For April babies, sweet pea is your birth flower.

During the Victorian Era (1837–1901), the language of flowers associated sweet peas with pleasure and departure. A farewell gift of

sweet peas to one's host meant "Thank you for a lovely time." During the era's time of conservative etiquette, a lady could scandalously signal "Meet me" to her lover by wearing a few sweet peas on her person. Now, parents give sweet peas to their daughters on their wedding days, symbolizing bittersweet farewell.

We use "sweet pea" as a term of endearment for children, pets, and romantic partners, comparable to "honey bunch" or "sweetie pie." Its use as an affectionate nickname reflects its fragrance and beauty.

APRIL 12

Nature as Inspiration

> A caterpillar spits out a sac of silk
> where it lies entombed while its genes
> switch on and off like lights
> on a pinball machine. If every cell
> contains the entire sequence
> constituting what or who the creature is,
> how does a certain clump of cells
> know to line up side by side
> and turn into wings, then shut off
> while another clump blinks on
> spilling pigment into the creature's
> emerald green blood, waves of color
> flowing into wingscales—black, orange,
> white—each zone receptive only to the color
> it's destined to become. . . .
> —ALISON HAWTHORNE DEMING, *The Monarchs:*
>   *A Poem Sequence*

Science seeks to discover and explain facts about the natural world. Poetry explores thoughts, feelings, experiences, sensations, and emotions. Rather than opposed, science and poetry are complementary ways of perceiving and interpreting nature. Natural history falls in between science and poetry in subjectivity, because observations of nature are filtered through the observer's knowledge, prior experiences, and value and belief systems. All three—the scientist, naturalist, and poet—focus on the patterns and processes of nature.

April is National Poetry Month in the United States and Canada, inaugurated in 1996 by the Academy of American Poets. Why not use National Poetry Month as an invitation to enjoy poetry by Diane Ackerman, Alison Deming, Ralph Waldo Emerson, Robert Frost, Henry David Thoreau, William Wordsworth, and other poets who found inspiration in nature?

APRIL 13

Bundles of Infinite Energy

April sees the month-long Alaska Hummingbird Festival in Ketchikan, Alaska, the southeastern-most city in the state. The festival welcomes back returning rufous hummingbirds from the southern US and Central America, a round-trip migration that can encompass up to 12,000 miles (19,312 km). By mid-March, rufous hummingbirds begin arriving, and by mid-April, they are seen throughout Ketchikan, slurping nectar from flowers and feeders. The festival affords an opportunity to watch, draw, and photograph these seemingly joyous beings in constant motion—hovering, flying backward, and dancing on air.

Hummingbirds have delighted and amazed us for centuries. When Europeans first saw hummingbirds, they imagined these iridescent New World jewels to be a cross between large buzzing insects and miniature birds. We attribute hummingbirds with courage, as many species are territorial and aggressively defend their flower patches. They chase other hummers and sometimes engage in fights, though generally an individual loses no more than a feather or two. As anyone with hummingbird feeders knows, sometimes one bird will guard the feeder and refuse to share with other sugar addicts. We are impressed with their endurance and perseverance. Little 3- to 5-inch (7.6- to 12.7-cm) ruby-throated hummingbirds breed in North America, but every autumn they fly to southern Mexico or Central America to overwinter. Some individuals fly nonstop up to 500 miles (805 km) over the Gulf of Mexico—18 to 22 continual flying hours. Hummingbirds symbolize "accomplishing that which is seemingly impossible."

APRIL 14

Helpers' High

Why do some animals help one another? Prairie dogs warn others of danger with calls that, paradoxically, might attract attention to themselves. Vampire bats regurgitate blood meals to hungry bats in their social groups. Because these bats can starve within 2–3 days without feeding, the donation is crucial for those individuals unlucky in foraging. The donor, however, might compromise its own health. For many species, altruism (behavior that benefits the recipient at a potential cost to the helper) occurs more frequently between genetically related individuals. Thus, the behavior benefits the genes of the helping individual as well as the recipient. Some species exhibit reciprocal altruism—you scratch my back, and I'll scratch yours.

April is National Volunteer Month in the United States. The World Giving Index, published by the Charities Aid Foundation since 2010, uses Gallup surveys to gather information on three charitable actions by respondents within the previous month: donating money, volunteering time, and helping a stranger. A country's index is the average of the three percentages. For the 2017 index, over 146,000 people were surveyed from 139 countries (on average more than 1000 people/country). Topping the list of the most "giving" countries were Myanmar (65 percent), Indonesia (60 percent), and Kenya (60 percent).

Helping each other seems to be part of the human spirit. An optimistic interpretation is that because human society is based on cooperation, helping may be a natural instinct. A more pessimistic interpretation is that we help because society expects it of us. Either way, volunteers who give of their time, energy, and talents enjoy a "helpers' high."

APRIL 15

Cat Goddess

Each year, thousands of ancient Egyptians traveled by boat to Bubastis in the southeast Nile Delta on April 15, on a pilgrimage to honor the cat goddess Bast (Bastet). The travelers partied on the river with music, dancing, and singing. Once in Bubastis, they feasted and drank more wine during the festival than during the rest of the year combined, for the goddess was to be celebrated with drunkenness. This was the Festival of Bast (also known as the Festival of Intoxication). According to the Greek historian Herodotus (ca. 484 to ca. 425 BCE), the Festival of Bast was one of the most joyous of all ancient Egyptian festivals, attracting as many as 700,000 revelers.

In the third millennium BCE, Bast was depicted either as a fierce lioness or as a woman with a lioness head. As such, she was a warrior goddess and protectress of Lower Egypt. Sometime between 945 and 715 BCE, the goddess evolved from a lioness to a domestic cat deity: a cat-headed woman. In that capacity, she had the playfulness, affection, and cunning of a domestic cat—but still the power of a lioness. Priests kept sacred cats, regarded as incarnations of the goddess, in the Temple of Bast. When each cat died, it was mummified and offered to Bast. When the temple in Bubastis was excavated in 1887–1889, more than 300,000 mummified cats were discovered. The cache reflects the reverence ancient Egyptians held for cats, and for their cat goddess.

APRIL 16

Illustrations Matter

A picture really is worth a thousand words, especially in a children's book. April 16, 1912, saw the birth in New York City of children's book illustrator Garth Williams. Williams' parents were both full-time artists. One day when Williams was a toddler, his dad watched him draw a tree on a window steamed with condensation. There was no doubt—this kid had talent. In time, Williams illustrated nearly 100 books, many of which became classics in American children's literature.

Generations of kids have adored the animals lovingly portrayed by Williams in *Stuart Little*, *Charlotte's Web*, *Bedtime for Frances*, *Little Fur Family*, and *Wait Till the Moon Is Full*, among others. Mice, pigs, raccoons, rabbits, bears, kittens, crickets, and spiders become friends. Williams claimed he used his illustrations to awaken children's interest in the world, sense of humor, and appreciation of life.

In 2012, sociologist J. Allen Williams and his colleagues published a study in which they examined illustrations in the 296 Caldecott Prize-winning medal and honor children's books from 1938 to 2008. They reported a steady decline in nature images (wild animals and the natural environment) and an increase in "built environment" images (human constructs, such as homes). The sociologists speculated that this trend reflects our increased isolation from nature, caused by urbanization, television, and other factors. The world needs more passionate artists like Garth Williams to encourage children to embrace nature.

APRIL 17

America's Black Bear

The Bayou Teche Black Bear Festival is held in Franklin, Louisiana, the third weekend of April. At one time the local Chitimacha Native Americans relied on Louisiana black bears for their fur, meat, and fat and honored them in cultural and religious traditions. With the introduction of modern firearms, the bears declined. Now, thanks to protective measures, the bears have rebounded. The festival provides an opportunity to learn about Louisiana black bears while enjoying music, food, and field trips. The organizers suggest bringing your child's favorite teddy bear along.

Why would one bring a teddy bear to the Bayou Teche Black Bear Festival? A black bear served as the inspiration for the first teddy bear, named after President Theodore "Teddy" Roosevelt. In

November 1902, Roosevelt participated in an American black bear hunt in Mississippi. Several members of the hunting party had already killed bears; Roosevelt had not. Not wanting the President to be left out, several hunters tracked down a black bear, clubbed it, and tied it to a willow tree so the President would have a bear to shoot. When Roosevelt saw the animal he refused to shoot it, declaring it unsportsmanlike. He ordered someone else to put the bear out of its misery. The story got out, and Roosevelt's sympathy for the injured bear inspired toymakers Morris Michtorn (US) and Richard Steiff (Germany) to create the first teddy bears in 1903.

APRIL 18

Pennies for Pandas

Giant pandas depend on bamboo leaves for 99 percent of their diet. In an ecologically perfect world, pandas would migrate to a better spot when bamboo disappears from their forest. But pandas' forests are shrinking because of habitat destruction and fragmentation, and the bears often have nowhere to move. The problem is compounded because bamboo plants flower after a growth cycle of anywhere from three to over 100 years. Most species of bamboo undergo "mass flowering," during which all plants in a cohort flower within a period of several years or less. Giant pandas won't eat bamboo once it flowers. After a plant flowers, it produces seeds and dies. Seeds take many years to germinate. Thus, regeneration of a bamboo forest is slow despite rapid post-germination growth. In the early 1980s, bamboo in southwestern China started to flower, and giant pandas began to starve.

US First Lady Nancy Reagan served as the honorary national chairwoman for a "Pennies for Pandas" fund-raising campaign through the World Wildlife Fund. She asked schoolchildren to donate pennies to provide food for starving pandas in China. On April 18, 1984, Mrs. Reagan received a check for nearly $10,000, all of it donated by children. The following week, she and President Reagan delivered the check and two jeeps to the China Wildlife Conservation Association when they visited the country.

Ongoing conservation efforts have paid off. Surveys of wild pandas during 1985–1988 estimated 1114 individuals; in 2004, 1596; and the latest census (2014), 1864. In September 2016, the giant panda was moved from the International Union for Conservation of Nature and Natural Resources (IUCN) Endangered category to Vulnerable.

APRIL 19

Saved from Extinction

The California condor, with a wingspan of 9.5 feet (2.9 m), is the largest bird in North America and one of the rarest worldwide. Although these majestic birds once soared over much of North America, their populations had dwindled dramatically by the 1950s due to habitat destruction, poaching, electric power lines, and lead poisoning from eating carcasses containing lead shot.

By 1982, only 23 California condors remained. To save the condor from extinction, biologists initiated captive breeding programs at the San Diego Zoo and the Los Angeles Zoo. On April 19, 1987, the last remaining California condor was removed from the wild. Birds in the two zoos laid eggs, and periodically captive-bred condors were released into the wild. By 2008, more California condors soared free than remained in the captive programs. By the end of 2016, 276 condors lived in the wild and 170 in captivity.

For the Wiyot people from California, the condor symbolizes human rebirth, as reflected in one of their sacred stories. Gouriqhdat Gaqilh, creator of all things, became angry with the wicked ways of men and women and sent a flood to drown all of Earth. Condor, the sole survivor, created a new world—one that included humans cleansed of evil. In a twist of fate for the Wiyot, having been saved

by Condor, the condor itself has been reborn through human intervention . . . at least for now. The California condor is still listed as Critically Endangered, largely because of ongoing lead poisoning from carcasses it scavenges.

APRIL 20

Fish on Fridays

In the early 1960s, Lou Groen owned a McDonald's franchise in Cincinnati, Ohio. Groen's restaurant was located in a neighborhood that was 87 percent Catholic. On Fridays he sold so few hamburgers that he took in about $75. Thus, the idea of a fish burger was born. Groen pitched his idea to the company owner, who had his own idea of a nonmeat burger: the Hula burger, a ring of grilled pineapple topped with melted cheese on a bun. The owner made a deal—they would sell the two nonmeat sandwiches at selected locations on Good Friday, April 20, 1962. Whichever sold more would be added to the McDonald's menu. The score at the end of the day was Hula 6, Filet-O-Fish 350. Filet-O-Fish became a McDonald's staple in 1965.

In January 2013, McDonald's USA announced it would display the Marine Stewardship Council's (MSC) blue ecolabel of sustainability on its fish packaging in restaurants nationwide, meaning that the fish (wild-caught Alaska pollock) can be traced back to a fishery that meets the MSC's strict standards of sustainable fishing practices. For a fishery to become MSC-certified, it must operate at a level such that fishing activity can continue indefinitely; the structure, productivity, function, and diversity of the ecosystem must be maintained for minimal environmental impact; all relevant laws must be followed; and the management system must be effective and responsive to changing circumstances. With more than 300 million Filet-O-Fish sandwiches sold annually, McDonald's use of sustainably caught Alaska pollock sends a powerful message to its patrons.

APRIL 21

Suckled by a Wolf

April 21, 753 BCE, is the traditional date given for the founding of Rome. According to legend, Amulius usurped the throne of his older brother Numitor, King of Alba Longa. To ensure no male heirs, Amulius killed Numitor's sons and forced the daughter, Rhea, to become a vestal virgin. In time, either having betrayed her vows of chastity or having been raped, Rhea gave birth to twin boys, Romulus and Remus. The alleged father was Mars, god of war.

Amulius imprisoned Rhea and ordered a servant to throw the twins into the Tiber River. Instead, the servant abandoned the basket with the twins on the riverbank. The Tiber rose and swept the twins downstream until the basket lodged in the roots of a large tree. A she-wolf found and suckled the boys, and a woodpecker brought them solid food. In time, a shepherd discovered and raised the boys. Once he reached early adulthood, Romulus killed Amulius, and the twins were rewarded with the throne of Alba Longa. Instead of ac-

cepting, they restored Numitor to the throne and set about to establish their own city. The 18-year-old twins quarreled over the location for their new city. Romulus favored the Palatine Hill, Remus the Aventine Hill. Romulus killed Remus, named the city Roma after himself, and declared himself King.

Were Romulus and Remus real people? Historians still argue this point. How likely is it that a she-wolf would suckle human infants? About zero, but the legend reflects the positive perception of wolves by the ancient Romans who believed the wolf was sacred to Mars.

APRIL 22

Photography and Wilderness Preservation

> You don't take a photograph, you make it.
> —ANSEL ADAMS, photographer

Naturalists, conservationists, environmentalists, and others who dedicate their lives to preserving landscapes strive to accomplish their goals in different ways depending on their individual talents. Ansel Adams, a pioneer of landscape photography, helped preserve some of the US West's most spectacular places through his stunning black-and-white photographs that captured the magnificence and mystique of wilderness. The popularity of his photography allowed him to garner support for his relentless efforts to protect wilderness.

Adams grew up in the still-wild reaches of the Golden Gate, San Francisco, California, in the early 1900s. As a child, he roamed the sand dunes and surrounding areas. Once he began exploring the Sierra Nevada Mountains, Adams knew he had found his spot on Earth. He hiked, climbed, and began recording his surroundings with the Kodak No. 1 Brownie box camera his parents had given to him at age 13.

In 1980, President Jimmy Carter awarded Adams with the Presidential Medal of Freedom, the nation's highest civilian honor. The citation read: "It is through [Adams'] foresight and fortitude that so much of America has been saved for future Americans." Adams died on April 22 (Earth Day), 1984. The following year, a mountain peak in Yosemite National Park was officially named Mount Ansel Adams, a proper tribute to the man who loved the majestic Sierra Nevadas.

APRIL 23

Free to Explore

> In the end we will conserve only what we love, and we will love only what we understand, and we will understand only what we are taught.
> —BABA DIOUM, Senegalese environmentalist (words spoken at the IUCN General Assembly, New Delhi, 1968)

National Environmental Education Week is held each mid-April. Environmental education got a boost in September 2008, when the US Congress passed the No Child Left Inside (NCLI) Act. The NCLI included funding environmental education at the national level. As of 2017, over 2000 schools, museums, zoos, botanical gardens, and other organizations have joined the NCLI coalition to provide kids with a strong foundation for understanding environmental issues and the skills they will need to make informed decisions in the future.

The great outdoors offers the perfect classroom to teach environmental education. In the 1950s, "outdoor preschools" (schools where children spend most of the time outside) became popular in Sweden and Denmark. By the 1960s the concept had spread to Germany and later to the United Kingdom. The idea has since crossed the ocean to North America and spread elsewhere. Outdoor preschools encourage children to observe, question, explore, and bond with nature. Proponents of outdoor classrooms assert that children educated outdoors retain information better than those educated indoors, exhibit fewer behavioral problems, learn self-reliance and adaptability, and develop a love of learning. For children who go to traditional schools, every schoolyard, whether inner-city or rural, provides opportunities for inquiries that encourage children to think and learn—and ultimately want to conserve nature.

APRIL 24

Respect for Life

Jainism, an ancient Indian religion, teaches nonviolence and non-injury toward all living beings. All animals are sacred, even the rats that eat grain and steal human profits and the mosquitoes that transmit disease. Jains often brush the path to avoid stepping on ants or other small invertebrates. There are about 4 million Jains worldwide, most of whom live in northwest India. Instead of belief in a monotheistic God, Jains look to 24 Tirthankars, persons who have achieved enlightenment while on Earth. Tirthankars serve as human spiritual guides to point the way to enlightenment (nirvana) for those seeking to liberate their souls from the cycle of birth and death.

The Jains' most important religious festival is Mahavir Jayanti, commemorating the birth in the sixth century BCE of Mahavira, the 24th and last Tirthankar. Mahavir Jayanti is celebrated on the 13th day of the waxing (rising) half of the Hindu month of Caitra, which is usually in April in the Gregorian calendar. Mahavira revived the Jain religion as it is practiced today. He was born a prince but left his kingdom and sought wisdom, enlightenment, and liberation in the forest. There, he determined that all life is sacred. As part of the Mahavir Jayanti festival, Jains donate to charitable causes, such as saving cattle from slaughter. Regardless of religious convictions, today is a good time for all of us to reaffirm our respect for life.

APRIL 25

Self-Important Little People

"Let's watch the video about jungle penguins," suggested my five-year-old granddaughter, Fionna. I explained that penguins don't live in the jungle and that her video must have featured some other black and white bird. She insisted it was a penguin. Long story short, I had to eat my words. The *tawaki* (Fiordland crested penguin) from South Island, New Zealand, is unique among penguins—they build their nests in lush temperate rainforest.

Polls worldwide reveal that penguins are many people's favorite birds. The 17 species of penguins range in size from the little blue fairy penguin, which stands about 16 inches (40 cm) tall, to the emperor penguin, at 3.6 feet (1.1 m). We delight in their quirky antics, adorable looks, and goofy comportment, as they waddle about like self-important little people.

Today is World Penguin Day. The celebration began at McMurdo station in Antarctica, where researchers noticed that every year about April 25 hundreds of Adele penguins returned to the same spot for their winter foraging ground. Each year, the researchers gathered on shore to welcome back the penguins on the day they called Penguin Day. Eventually the day became known as World Penguin Day, a day that invites people everywhere to celebrate penguins and to support their conservation. Join in—don your black and white clothes, waddle through the day, and eat fish sashimi; watch *March of the Penguins*; or greet the penguins at your local zoo or aquarium.

APRIL 26

Gold Standard

Today is National Audubon Day. Famed bird artist John James

Audubon was born on April 26, 1785, in the French colony of Saint-Domingue (now Haiti), the son of a French sea captain/adventurer/plantation owner and his Creole mistress. His mother died soon after his birth, and his stepmother raised him in France. Audubon was passionate about birds, music, drawing, and painting, but he was a spoiled kid who resisted education and discipline. At the age of 18, Audubon's father sent him to the family estate near Philadelphia, Pennsylvania, US, largely to avoid conscription in Napoleon's army. There, Audubon passed his days fishing, hunting, and drawing birds.

Audubon later spent more than a decade involved in business ventures ranging from a dry-goods store to a lumber mill, all of which failed. Throughout this time he continued to draw birds. In 1819, married, with two sons, and not much more than his gun and bird drawings, he focused on doing what he did best—painting birds, with the goal of painting all the birds of North America. In 1824, he sought a publisher in Philadelphia, without luck. Audubon sailed to England in 1826 to secure a publisher for his growing collection of more than 300 watercolors. Europeans, enamored with natural history during the Romantic Era, adored his work. Audubon secured a publisher in Edinburgh and London for what he would call *Birds of America*, a collection of 435 prints. The last print was issued in 1838. Audubon's exquisite paintings are still the standard against which other avian art is measured.

APRIL 27

Save the Frogs

Populations of frogs are experiencing significant declines and extinctions worldwide, with 32 percent of species identified as threatened with extinction. Frogs are essential components of food webs, both as prey and predators. If one-third of the world's frogs disappear, diverse ecosystems will suffer. But do we care? A 2010 Gallup poll revealed that only 31 percent of over 1000 US adults questioned were concerned about extinction of plants and animals. To combat this apathy, various conservation organizations are focusing more attention on public education.

One such organization is Save the Frogs, founded by ecologist Kerry Kriger in 2008. This international conservation organization is composed of scientists, naturalists, educators, policymakers, and laypersons. Its stated mission is "to protect amphibian populations and to promote a society that respects and appreciates nature and wildlife." The organization's primary educational event is Save the Frogs Day, celebrated each year on the last Saturday in April. Between its founding in 2008 and 2017, Save the Frogs has sponsored more than 1900 educational programs, held in 57 countries.

Frogs symbolize happiness, good luck, fertility, and love in both Western and Eastern folklore. People worldwide carry frog-shaped amulets, charms, and talismans to protect against injury, disease, and death and to attract good fortune, happiness, and health. How incongruous that an animal that has long protected and improved our lives should now be facing its own demise!

APRIL 28

Monster Medicine

Gila monsters spend about 95 percent of the year underground or in shelters. They fast for months at a time, and when they eat they can consume more than one-third their body mass. How can these lumbering lizards deal with this much food at once? Scientists identified a hormone—exendin-4—in Gila monster saliva that is released when the lizard bites into its long-awaited meal, apparently priming the gastrointestinal tract.

In humans, the hormone insulin regulates the body's use of sugar and other food. Glucose builds up in the blood without the correct amount of insulin (type 1 diabetes) or if the body does not use insulin properly, a condition called insulin resistance (type 2 diabetes). Physiologists identified a hormone they called GLP-1 that is secreted from the walls of our intestines when we eat. GLP-1 stimulates release of insulin from the pancreas. Researchers asked: If you gave diabetics GLP-1, would it lower their glucose levels? Yes, but the hormone lasts only a few minutes in the body before an enzyme breaks it down.

Once medical researchers learned about exendin-4 in Gila monsters, they found the hormone to be an analogue of GLP-1. The good news is that exendin-4 is extremely resistant to being broken down by the enzyme that breaks down GLP-1. Eureka! On April 28, 2005, Exenatide (marketed as Byetta), a synthetic version of exendin-4, known only from Gila monster saliva, was approved by the FDA and is now used to treat type 2 diabetes.

APRIL 29

Circumventing and Hijacking Toxins

On April 29, 1996, three California chefs were hospitalized (but survived) after eating 0.25 to 1.5 oz of contaminated *fugu* (pufferfish) brought from Japan by a coworker as a "prepackaged, ready-to-eat product." *Fugu* is a risky delicacy. Bacteria associated with the fish produce tetrodotoxin, a potent neurotoxin that renders pufferfish poisonous to predators. Japanese *fugu* chefs must complete three or more years of rigorous training before being certified to prepare the fish for consumption. The most toxic body parts (liver, ovaries, and eyes) must be removed without contaminating the flesh.

The natural world is awash in poisons. Some plants, mushrooms, and bacteria produce strychnine, amatoxin, and botulinum, respectively. Animals use a rich array of toxins to defend themselves against predators, but, like the pufferfish, they don't always manufacture it themselves. Poison frogs (family Dendrobatidae) store defensive chemicals in granular skin glands and secrete these tox-

ins when attacked. The frogs get their toxins from the ants, mites, beetles, and millipedes they eat. Humans hijack the poisons from these frogs. In western Colombia, native hunters tip their blowgun darts with the toxic secretions from three species of poison frogs to kill game, including deer, bear, and jaguar.

Toxins don't always protect animals from the clever predator *Homo sapiens*. But imagine the trial and error required before people could use frog toxins to enhance weapons and could eat pufferfish safely!

APRIL 30

Akhal-Teke

The last Sunday in April is Turkmen Horse Day in Turkmenistan, an annual racehorse day. Why would you care about this holiday if you're not a Turkmen? Because of the horses themselves. The racehorses are Akhal-Teke, one of the oldest existing horse breeds. For thousands of years these horses have been prized for their speed, stamina, intelligence, and beauty. After Turkmenistan gained its independence when the Soviet Union dissolved in 1991, these horses rose to the status of national emblem and now are featured on the coat of arms, stamps, and banknotes. The choice of Akhal-Teke to

represent the Turkmen spirit is fitting, since horseracing is the nation's favorite sport.

Although the ancestry of Akhal-Teke is difficult to trace, some experts believe they are descended from horses ridden by Mongol raiders during the thirteenth and fourteenth centuries. These horses are tough and resilient, having adapted to harsh desert landscapes. They can cover long distances with minimal rest, food, and water. Akhal-Teke served tribal inhabitants of what is now Turkmenistan during their near-constant intertribal wars and during their raids on neighboring Persia.

Now the Akhal-Teke of Turkmenistan are used for racing, endurance riding, and show jumping. Mass gatherings in Turkmenistan generally are permitted only to extol the state and its leaders. Horseracing, however, is an exception. No wonder, then, that the annual Turkmen Horse Day is a cherished event.

MAY 1

Arrival of Spring

Today is May Day, a celebration that dates back to the Druids of Ireland, Scotland, and the Isle of Man with the festival of Beltane.

Bonfires were set on this day to enhance the protective power of the sun. Homes and cattle were decorated with yellow flowers, and people feasted and drank. When the Romans occupied the British Isles, they brought with them worship of Flora, goddess of flowers, fruits, vegetables, and fertility. They celebrated the festival of Floralia, in honor of Flora, from April 28 to May 3 in hopes of pleasing the goddess so she would protect the blossoms and ensure fertility. Festival-goers wore floral wreaths in their hair. Prostitutes danced naked, and participants played games and watched theatrical performances. In time, the rituals of Beltane and Floralia merged in the British Isles.

May Day is still celebrated in Europe and North America to welcome the arrival of spring. Customs reflect old European traditions, and flowers nearly always play a central role. When I was a child growing up in western Pennsylvania, my friends and I followed the local custom: we made up baskets of wildflowers, rang our neighbors' doorbells, left the baskets, and ran before being seen. In England and parts of Germany, a tall pole is decorated with flowers; children grab the ends of long ribbons attached to the top of the pole and sing and dance around the "maypole." In Greece, people hang wreaths of wildflowers on their balconies or front doors. Today, as in those early celebrations of Beltane, flowers represent new beginnings.

MAY 2

Gardening au Naturel

> Would that you could meet the sun and the wind with more of your skin and less of your raiment, for the breath of life is in the sunlight and the hand of life is in the wind. . . . And forget not that the earth delights to feel your bare feet and the winds long to play with your hair.
> —KHALIL GIBRAN, *The Prophet* ("On Clothes")

The first Saturday of May is World Naked Gardening Day, begun in Seattle, Washington, in 2005. The event has spread worldwide, especially in the UK and elsewhere in Europe. It's a day to spend a few hours in the buff, hoeing the beds, planting flowers or vegetables, weeding, or trimming trees and shrubs. Experienced naked gardeners suggest there's no better way to celebrate and enjoy nature than the way nature intended us to be—in our birthday suits. Enthusiastic greenhorns claim that the feeling of a soft breeze and warm sunshine on exposed skin is transforming once they toss off self-consciousness and concern over getting arrested for public indecency.

Words of warning: Check into the legality of nude gardening in your area. In some communities, if the neighbors see you and call the police, you could be slapped with a fine. And watch out for the roses, briars, nettles, and poison ivy—and mosquitoes, biting flies, fire ants, chiggers, and ticks!

MAY 3

Animals in War

On May 3, 2010, it was announced that Steven Spielberg would direct the film *War Horse*, a story about a horse trained for military operations during World War I. About 10 million soldiers died during that war, but less well known is that 8 million military horses died as well.

Animals have long been used in war. In 190 BCE, the great Carthaginian general Hannibal catapulted earthenware jars full of venomous snakes onto the ship of Eumenes, king of Pergamum. The king and his men retreated in panic and terror, and Hannibal won the battle despite being greatly outnumbered in ships. During World War I, bioluminescent light from European glowworms housed in jars allowed soldiers in dark trenches to study maps, examine intelligence reports, and read letters from home. Camels carried soldiers and supplies during combat operations in the Middle East and North Africa during both World Wars. During the Vietnam War, Viet Cong soldiers placed pit vipers in underground bunkers or along trails as living booby traps. In Vietnam and Iraq, dogs detected mines, sniffed out weapons, and rescued injured soldiers. Using their amazing echolocation skills, bottlenose dolphins detected and marked underwater mines during the Vietnam and Persian Gulf Wars.

We use animals during warfare to protect ourselves because of their superior senses (dogs, dolphins), strength (camels, horses), lethality (venomous snakes), or other attributes (glowworms). The inherent, and debatable, assumption is that a nonhuman animal's life is more expendable, less precious, than a human life.

MAY 4

Cage-Free

We depend on chickens as a source of meat and eggs, yet we often abuse them. In 2005, United Poultry Concerns, an organization that promotes the compassionate and respectful treatment of domestic fowl, launched a day to raise awareness of chickens' suffering worldwide. May 4 is International Respect for Chickens Day.

Until recently, most egg-laying hens in the US have been confined to cages in which the average space per hen is less than the size of an 8 × 10-inch sheet of paper—for their entire adult lives. Many egg producers are switching to an indoor "cage-free" system in which the hens can walk, spread their wings, and lay eggs in nests. Some producers use a "free-range" system in which hens have continuous access to the out-of-doors during their egg-laying cycle. "Pasture-raised," an unregulated label, refers to hens that roam free and graze in a pasture.

Many restaurants and supermarkets have pledged to use and sell only cage-free eggs. By the end of 2016, Taco Bell offered only cage-free eggs on their breakfast menu. Burger King vowed to complete a switchover to cage-free eggs by 2017; Starbucks and Wendy's by 2020; and McDonald's, Subway, Dunkin' Donuts, Trader Joe's, Wal-Mart, and Target by 2025. Cage-free eggs will cost the consumer more, but think of the good karma accrued for producers and consumers alike. Perhaps someday people will look back and shudder at their ancestors' barbaric exploitation of chickens.

MAY 5

Animal Companionship

The first full week of May is National Pet Week. According to the 2017–2018 National Pet Owners' Survey, 68 percent of US households (85 million families) own at least one pet. Whether the pet is a dog or cat, horse or llama, hamster or guinea pig, snake or lizard, scorpion or tarantula, guppy or angelfish, we value our pets for the social and emotional needs they fulfill. When we vacation, we can leave our dogs in luxury pet resorts where they can swim, walk on the beach at sunrise and sunset, and receive a massage, mud bath, or "pawicure." Our cats can enjoy suites with climbing shelves, bird feeders outside their windows, skylights, and a pureed chicken snack before evening bedtime. Pets are so much a part of us that many US states provide sections of cemeteries where people can be buried next to their beloved pets. The ancient Egyptians set a precedent for this. Pet dogs, cats, and monkeys were mummified and buried in tombs with their owners.

Today, if you live in the congested metropolis of Tokyo, your landlord likely forbids pets in your tiny apartment. Ergo, "animal cafes" have become popular in Tokyo, where customers pay to hold and pet animals. First there were cat cafes, then rabbit and owl cafes. A hedgehog cafe opened in 2016: 1000 yen ($9.20) will get you 30 minutes of holding time on a weekday. On the weekends, the charge is 1300 yen for 30 minutes. If you live where you can have a pet, rejoice.

MAY 6

Fairy Cups

> White coral bells upon a slender stalk
> Lilies of the valley deck my garden walk.
> Oh, don't you wish that you might hear them ring?
> That will happen only when the fairies sing.
> —Traditional children's song

Folklore sometimes refers to lilies of the valley as "fairy cups," imagined as little white cups fairies use for drinking fairy-wine, hung on slender stalks while the sprites dance. Other lore tells that the little bell-shaped flowers ring out whenever fairies sing, or that the flowers serve as ladders for fairies to reach the reeds they gather for weaving their cradles.

Lilies of the valley are also called "Mary's tears," from the legend that when Mary cried at Jesus's crucifixion, her tears turned

into white, cup-shaped flowers. Another legend says the flowers originated from Eve's tears after she and Adam were expelled from the Garden of Eden. French legend tells of a holy man named Saint Leonard who went to live alone in the woods to commune with God. There he encountered the dragon Temptation. Bloody battles ensued, and wherever dragon blood touched the ground, poisonous weeds sprang up. Saint Leonard's spilled blood yielded lilies of the valley.

These "white coral bells" are associated with happiness, sweetness, purity, and humility. During the Victorian Era (1837–1901), a gift of these flowers conveyed the message: "You have made my life complete." Lilies of the valley are the birth flower for May.

MAY 7

Elegant Irritations

A tiny parasite becomes trapped inside a mollusk's shell. The mollusk's immune system responds by creating a "pearl sac" from cells and calcium carbonate secretions to seal off the irritant. The many layers formed around the parasite produce a pearl. Natural gem-quality pearls are hard to find. Three tons of pearl oysters might yield three or four perfect pearls.

On May 7, 1934, a Filipino diver found the world's largest pearl (14.1 pounds; 6.4 kg), the Pearl of Lao Tzu (also known as Pearl of Allah), off the Philippine island of Palawan. Found inside a giant clam, it is not considered a gemstone pearl, but rather a "clam pearl." Pearls that form inside saltwater pearl oysters and freshwater pearl mussels have iridescence. In contrast, pearls that form inside giant clams have a porcelain-like surface. Nonetheless, this giant pearl is valuable. In 2007, a gemologist appraised its value at $93 million.

Natural pearls are rare jewels, their value based on luster, size, color, lack of surface flaw, and symmetry. Pearls were symbols of

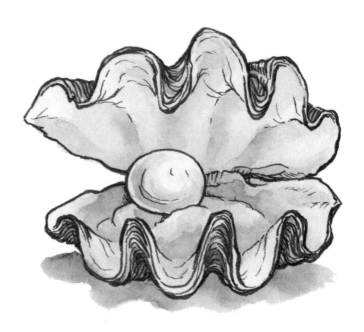

wealth, class, and status for the ancient Greeks and Romans. Cleopatra (69–30 BCE), the last Egyptian queen, wore pearl earrings. The Roman emperor Caligula (12–41 CE) adorned his favorite horse with a pearl necklace. Queen Elizabeth II's Imperial State Crown is decorated with 273 pearls. Richard Burton gave one of the world's most famous pearls, La Peregrina, to his wife, Elizabeth Taylor, for Valentine's Day in 1969. An African slave found La Peregrina, a large, perfectly symmetrical, pear-shaped pearl, in the Gulf of Panama during the mid-sixteenth century. In return, the slave was awarded his freedom. What greater value could a pearl have?

MAY 8

Turtle-Champion

> Where twenty years ago most Caribbean shore was wilderness
> or lonesome cocal [coconut grove], aluminum roofing now
> shines in new clearings in the seaside scrub. The people are

breeding too fast for the turtles. The drain on nesting grounds is increasing by jumps. It is this drain that is hard to control, and it is this that will finish *Chelonia*.

—ARCHIE CARR, *The Windward Road*

In the mid-1950s, field biologist, ecologist, and conservation biologist Archie Fairly Carr Jr. warned of the impending decline of the green sea turtle (*Chelonia mydas*). Archie devoted his life to studying sea turtles, and through his research and writing he brought international attention to their plight. On May 8, 1987, Archie was awarded the Eminent Ecologist Award from the Ecological Society of America in recognition of "his outstanding research contributions, his efforts on behalf of the well-being of our planet, and his uncanny ability to communicate the excitement and the music of ecology to nonspecialists."

Archie's bond with sea turtles perhaps began when, as a young boy, he lay dozing on a warm dock, waiting for a fish to jiggle his line. A large loggerhead swam alongside the dock, thrust his head from the water, opened his mouth, and sighed. Archie awoke, and for a magical moment lay gazing into the sea turtle's eyes—an encounter that may have sparked Archie's lifelong concern for the future of turtles.

MAY 9

On the Move

Of the approximately 9930 species of birds, at least 2600 (26 percent) migrate from nesting to non-breeding grounds, or at least display substantial nomadism. The bar-tailed godwit sets the record for the longest known nonstop migratory flight—over 7000 miles (11,265 km), during nine days. Arctic terns, weighing about 3.5 ounces (100 g), fly the equivalent of the distance to the moon and back three times during their lifetime of about 30 years.

Many migratory birds face the challenge of unequal protection because they move across international borders. The Migratory Bird Treaty Act of 1918 (MBTA) implemented the 1916 Convention between the US and Great Britain (representing Canada) to protect migratory birds. Since then, the MBTA has implemented treaties between the US and Mexico, the US and Japan, and the US and the Soviet Union (now Russia). Without a waiver, it is unlawful to hunt, capture, kill, or sell birds on the MBTA list. The MBTA currently protects more than 800 species of birds.

But we need to do more. In 2015, Claire Runge and her colleagues reported in *Science* that half of all migratory species of birds are declining. Ninety-one percent of migratory bird species have inadequate coverage of protected area during at least one stage of their annual cycle; 18 species have no protected area coverage of their breeding grounds. Since 2006, the United Nations Environment Programme has sponsored World Migratory Bird Day on the second

weekend of May. The event raises awareness of bird migration and the international agreements, laws, and collaborations that protect migratory birds.

MAY 10

A Good Children's Story

> The mole had been working very hard all the morning, spring-cleaning his little home. First with brooms, then with dusters; then on ladders and steps and chairs, with a brush and a pail of whitewash; till he had dust in his throat and eyes, and splashes of whitewash all over his back fur, and an aching back and weary arms.
>
> —KENNETH GRAHAME, *The Wind in the Willows*

In the late spring of 1907 Kenneth Grahame, secretary of the Bank of England, and his wife spent a long holiday in Cornwall, England. Their 7-year-old son Alastair agreed to stay at home in London with his nanny *if* his father would continue telling bedtime stories by letter. On May 10, 1907, Grahame wrote a letter to Alastair that began *The Wind in the Willows*, the adventures of four anthropomorphized animals named Toad, Mole, Ratty, and Badger. Published in 1908, the book became one of the world's most beloved children's stories. It is more than an adventure story about four animals with wildly different personalities. On another level, it addresses everyday emotions that children experience: fear, nostalgia, awe, and wanderlust.

The year after *The Wind in the Willows* was published, US President Theodore Roosevelt wrote to Grahame and told him how much he enjoyed the book. As author C. S. Lewis once said, "A children's story that can only be enjoyed by children is not a good children's story in the slightest."

Maternal Sacrifice

> A mother's love for her child is like nothing else in the world.
> It knows no law, no pity, it dares all things and crushes down
> remorselessly all that stands in its path.
> —AGATHA CHRISTIE, "The Last Séance," in *The Hound of
> Death and Other Stories* (1933)

In the United States, Mother's Day is celebrated the second Sunday
of May—a time to honor mothers, motherhood, and the maternal
bond. As all mothers know, we give a large part of ourselves to our
children. Death is the ultimate maternal sacrifice, and some nonhuman animals routinely make such offerings.

A mother octopus aerates her hundreds or thousands of eggs by
shooting water through her siphon, removes debris from the egg
mass, and fights off octopus-egg-eaters, with little time to eat. She
never leaves her eggs and will likely die of starvation soon after her
young hatch. The babies abandon their empty egg capsules and their
dying mother, and swim off to begin life on their own.

A female Australian social spider hauls large insects for her
spiderlings to eat. She also sucks juices from these prey, storing fat
and developing additional eggs. With cooler weather, prey become
scarce. Nutrients from the developing eggs seep into her blood-
stream. The hungry young spiders attack their unresisting mother
and suck her nutrient-rich blood. Once she is dry, they inject venom
and devour the rest of her.

Be thankful you're human! Happy Mother's Day.

Cabinet of Curiosities

During the 1500s, Europeans began collecting and housing treasures in "cabinets of curiosities," also called "wonder rooms" because many consisted of entire rooms. These miniature museums featured wonders and oddities of the natural world, relics from antiquity, coins, scientific instruments, anything that struck the collector's fancy. One of the most famous collections during the early 1700s belonged to Albertus Seba, Dutch apothecary and collector extraordinaire, born on May 12, 1665.

Seba gathered plants and animals to prepare his remedies, but his real passion was collecting biodiversity. From his pharmacy near the international harbor in Amsterdam, Seba supplied departing sailors with medicines and in return asked them to bring him exotic plants and animals. When they returned to Amsterdam, he ministered to the sick and exhausted—and picked up his specimens. In 1717, Seba heard that Peter the Great intended to visit the Netherlands and shop for treasures for his own cabinet of curiosities. Seba sent the Russian tsar an itemized list of his collection, including 1000 insects, 72 drawers of shells, and 400 jars of animals preserved in alcohol. The tsar visited Seba and purchased the entire collection.

What does a collector do without his/her collection? Seba amassed an even larger one. Then, in 1731 he signed a contract with two publishers to produce a 4-volume work of over 400 plates illustrating his collection. Art and nature blend in what is known in English as Seba's *Thesaurus*, from shells arranged in geometric designs to lizards running across the page.

MAY 13

Windows to the World

May is Healthy Vision Month, established by the National Eye Institute in 2003. As the saying goes, "Eyes are our windows to the world." Vision allows us to appreciate the power of nature: electrical storms; ocean waves lapping, swelling, breaking, and crashing; the sun. The beauty of nature: colors and shapes of life; diversity of landscapes. The history of nature: potshards abandoned on the ground, remnants of past cultures; rolled-up trilobites exposed in road cuts, remnants of the Paleozoic. The behavior of nature: squirrels and chipmunks burying acorns; geese and cranes flying south. The magic of nature: mist forming over a lake; fog rolling into a pasture; the moon rising from behind a mountain peak.

Seven hundred million years ago, a group of marine invertebrates developed light receptors. In time, vision became critical for predators to locate prey and for prey to evade predators. Today, some of the most complex eyes are found in mantis shrimp. These marine crustaceans have 12 types of color receptors (we have only 3), and they have ultraviolet, infrared, and polarized light vision. The prize for the largest eyes goes to 50-foot (15 m) colossal squid, whose eyes measure about 12 inches (30 cm) in diameter—larger than a dinner plate! And then there are chameleon eyes: swiveling independently, able to see in two directions at once. As they say in Madagascar, "Like a chameleon, have one eye on the future, one eye on the past."

The US Camel Corps

Most camels today are domesticated. Camel milk is consumed fresh or made into yogurt or cheese; melted hump fat is spread like butter; high in protein and low in fat, meat is served as a delicacy. Wool is woven into warm coats and blankets; hides are stitched into fashionable shoes and saddles. Camel dung provides a slow-burning, odorless, and virtually smokeless cooking fuel. Most of all, we value camels for transportation and as beasts of burden.

Camels lived in North America millions of years ago. Could they live there again? Camels can survive longer without food and water than can horses, burros, and mules. In the mid-1800s, the US Army wondered whether camels could be used to build and traverse wagon roads to open up the southwest. On May 14, 1856, a shipment of 34

camels acquired from Algeria, Tunisia, Turkey, and Egypt was un-
loaded in Indianola, Texas. Nine months later, another 41 camels
were delivered.

The camels lived up to their reputation of hardiness, but the
army experiment was abandoned soon after the Civil War erupted.
Jefferson Davis was a chief sponsor of the experiment, and after
he became President of the Confederate States, Union supporters dropped the project. Some of the animals were sold at auction
and worked as pack animals in Nevada and Arizona, hauling wood,
ore, and supplies. Others were set free. Reports of the "last" free-
ranging camel in the American southwest range from 1901 to 1929.

MAY 15

Remove the Yuck Factor

By 2050, the world's human population is projected to reach 9.7
billion. The Food and Agricultural Organization of the United Nations estimates that to feed this number of people, we will need to
increase food production by 70 percent. Clearly, we must identify
alternative protein sources. Many nutritional experts suggest that
insects are the obvious alternative, with over 1900 species currently
recognized as edible. On May 14–17, 2014, Ede, Netherlands, hosted
the first "Insects to Feed the World Conference," to promote insects
as human food and as animal feed. Over 450 people from 45 countries attended.

More than 2 billion people, mostly in Africa, Asia, and Latin
America, already eat insects. Most eat insects whole: grasshopper
tacos (Mexico); fried queen ants (Brazil); raw honeypot ants (Australia); stir-fried cicadas (China); deep-fried silkworm pupae (Korea); boiled wasp larvae (Japan); fried giant water bugs (Thailand);
fried dragonflies (Indonesia); smoked caterpillars (Zambia); fried
cockroaches (Cameroun); and raw termites (Kenya). In contrast,
North Americans and Europeans generally consume their insects

disguised: cookies and energy bars made with cricket flour (Canada and US); tomato, carrot, and chocolate spreads containing ground-up mealworms (Belgium); pasta made with ground crickets and grasshoppers (France); and mealworm burgers (Netherlands). If we are serious about feeding the world with insects, and not just eating them as a novelty, we will need to remove the "yuck factor" for Westerners and convince people that eating insects is environmentally sustainable and that insects provide an excellent source of protein. Insects may be the food of the future.

MAY 16

Forest Deities

Imagine a forest fragment in India, left untouched. Now imagine 100,000 undisturbed fragments in forests, hills, and along rivers scattered across the country. These refuges exist because religious beliefs and cultural practices provide social fencing of "sacred groves" where nature is protected and revered. Deities associated with Hindu gods are believed to live in the trees and preside over the fragments, so it is taboo to cut down a tree. Sacred groves harbor rich biodiversity, and some serve as treasure troves for medicinal plants and the last refuge for rare plants and animals.

In the state of Manipur, northeastern India, these sacred groves are called *umanglai*, literally meaning "forest deities." The resident gods are believed to punish anyone offending an *umanglai* by causing illness or other misfortune. At the beginning of the wet season, usually late April or May, Manipuri hold week-long Lai-Haraoba ("festival of the gods") festivals at the 365 *umanglai* in the state to honor the deities. The gods are offered words, flowers, fruits, music, and dance with the hope that the deities will bestow prosperity on the village and protect the residents from sickness and harm. Festival rituals, including the songs and dances, reflect the people's love and respect for nature.

Now, imagine sacred groves scattered worldwide. Just think how different Earth would be if, as in Manipur where it is everyone's duty to protect the *umanglai*, people everywhere took responsibility for preserving nature.

MAY 17

The Heart of the Andes

> The river rapids in the centre of the painting were so realistic that people could almost feel the spray of the water. Trees, leaves and flowers were all rendered so accurately that botanists were able to identify them precisely, while the snow-capped mountains stood majestically in the background. More than any other painter Church had answered Humboldt's appeal to unite art and science.
> —ANDREA WULF, *The Invention of Nature*

American landscape painter Frederic Edwin Church (1826–1900) painted *The Heart of the Andes*, described above. Church's painting was first exhibited publicly, in New York City, between April 29 and May 23, 1859. Twelve thousand people paid a 25-cent entrance fee to view the oil-on-canvas 5-by-10 foot (1.5-by-3 m) majestic Ecuadorian landscape. Each of us responds uniquely to a painting. For me, the snow-capped Chimborazo volcano, jagged Andes, lush vegetation, and shimmering pool beneath a waterfall trigger a flood of sentimental memories from 1968, when, as a student, I experienced this landscape first-hand.

Church began by painting classic Hudson River School scenes in the northeast US, but after reading Alexander von Humboldt's book *Personal Narrative* about his exotic travels in Central and South America, Church was inspired to visit the area. Humboldt had challenged artists to portray the Andes in all their glory. Church visited twice and painted *The Andes of Ecuador, Cayambe, The Heart of the*

Andes, and *Cotopaxi*.

In early May 1859, Church wrote to a friend that he intended to send the canvas to Berlin to show Humboldt his painting. Sadly, Humboldt had died three days earlier, while *The Heart of the Andes* was being exhibited.

MAY 18

Nature's Resilience

On May 18, 1980, Mount St. Helens volcano in Washington State erupted, spewing hot rocks, ash, gas, and steam. The eruption impacted more than 230 square miles (596 sq km), causing the worst volcanic disaster in US history. Heat melted most of the glaciers and

snow on the mountain. Mudflows buried plants and animals, forests and meadows. Gray ash blanketed the landscape.

Mount St. Helens became a living laboratory for ecologists, an opportunity to watch survivors and colonizers reclaim devastated terrain. Certain plants re-sprouted from roots protected in the soil. Some animals survived the eruption, protected in underground burrows and crevices or under ice-covered lakes. Not all of these plants and animals tolerated the changed and extreme conditions, but some did. Wind carried new colonizers in the forms of fungal spores, seeds, spiders, and insects. In time, mammals and birds migrated into the impacted area. The eruption created ponds, springs, and other habitats needed by survivors and colonizers. Although many areas on the mountain still resemble barren deserts, most plants and animals previously found on Mount St. Helens are there today, nearly four decades later.

We often think of nature as fragile, but disturbance is natural. Change is continual. In order to survive, organisms must adapt to disturbance and change. The return of life on Mount St. Helens reminds us that nature is robust and resilient.

Life, Longevity, and the Turtle

Turtles symbolize longevity for cultures worldwide. Box turtles can live a century. Lonesome George, the last Pinta Island tortoise from the Galápagos Islands, is thought to have lived more than 100 years. A tortoise named Tu'i Malila holds the longevity record for a captive tortoise—188 years. In 1777 Captain James Cook presented Tu'i Malila, a radiated tortoise from Madagascar, to the royal family of Tonga. Tu'i Malila died on May 19, 1965.

I associate May 19 with life—the day in 1982 that my firstborn, Karen, came into the world. As my physician sister Cathy held my newborn, she shared that she had participated in both the begin-

ning and ending of life that day. She had watched a terminally ill patient die that morning, and by evening she was rejoicing in a new life.

Every May 19, as I recall Cathy's musing, I reflect on the cycle of life, longevity, and the turtle. In traditional Japanese wedding ceremonies, the bride might wear a kimono with tortoise images, and guests might be served tortoise-shaped cakes, symbolizing hope for a long marriage. I suppose turtle figures commonly appear on baby items, from disposable diapers to bibs, towels, and onesies, because they are adorable. But when I give a turtle gift to a newborn, I give it as a symbol of hope for longevity. When Cathy had her first child three years later, I gave him a ceramic wide-eyed baby turtle hatching from its eggshell.

MAY 20

Dürer's Rhinoceros

An Indian rhinoceros arrived in Lisbon, Portugal, on May 20, 1515, after four months at sea. Shipped from western India to King Manuel I of Portugal, it was the first living rhinoceros seen in Europe since Roman times. The animal served as the inspiration for German artist Albrecht Dürer's rhinoceros woodcut, though Dürer never saw it. Dürer based his woodcut on a sketch by an unknown artist who claimed to have seen the rhinoceros in Lisbon. The artist's sketch was not entirely accurate, leading Dürer to depict a rather fanciful rhinoceros with a covering of hard armor-like plates, scaly legs, a throat gorget, and a small twisted horn sprouting from its back.

The story of the Indian rhinoceros continues—briefly. King Manuel hoped to curry favor with Medici Pope Leo X to maintain papal grants for exclusive Portuguese possession of newly explored lands. Gifts were in order. King Manuel had already sent the Pope a white elephant from India. The rhinoceros seemed the perfect follow-up gift, so it was shipped off to Rome in December 1515. Unfortunately, it drowned en route during a shipwreck off the Italian coast.

Dürer's woodcut has inspired many other artists. It is represented on the door of the Cathedral of Pisa (Italy), in tapestries of Kronberg Castle (Taunus, Germany), in sculptures (including one by Salvador Dali), and in paintings. Dürer's rhinoceros will live on forever, even though it never met the Pope.

Untamed Nature

When Mary Reynolds, from County Wexford, Ireland, was five years old, she experienced plants whispering to her, asking to be nurtured. As a child, she drew designs for gardens based on nature. Later, she designed a stunning "Celtic Sanctuary," reflecting the value of wild places. Mary dreamed of entering her garden in the Chelsea Flower Show, an event sponsored by the Royal Horticultural Society, held in Chelsea, London, nearly every year since 1913. There, new plants are launched and older varieties are revived. Trees, flowers, vegetables, floral arrangements, and show gardens are exhibited.

Mary believed the only person who could build her "Celtic Sanctuary" for the show was Christy Collard, a builder with Future Forest Garden Center in West Cork, Ireland, whom she had met in gardening circles. He proved elusive, so she traveled to the parched highlands of Ethiopia, where he was building a tree garden. Collard returned with Mary, and he and the Future Forest crew built "Celtic Sanctuary," featuring a sheep pasture, 500 wild plants, and a stunning wall made from stone from Cork, Ireland.

May 21 was the first day of the 4-day Chelsea Flower Show in 2002. There, 28-year-old Mary won a gold medal for her garden, winning over those of world-class garden designers, as well as the "Healing Garden" entered by Prince Charles. The "Celtic Sanctuary" encouraged people to invite nature back into their gardens. The garden and a movie entitled *Dare to Be Wild*, released in Ireland in 2016, are a tribute to the whispering plants from Mary's youth.

Biodiversity

Imagine if only one species of bird visited our feeders throughout the winter. If we could plant only one type of flower in our gardens, and only one kind of butterfly slurped nectar from those flowers. If only one species of large mammal congregated at water holes across Africa. If only one kind of tropical fish inhabited the Great Barrier Reef. Would we cling to what was left of nature, or would we wither away like so many species before us?

Conservation biologists warn that we are in the early phase of Earth's sixth major extinction episode. This episode differs from the previous five. The extinction rate now is much greater than in the past, and our activities cause most current extinctions.

The international community has united to protect biodiversity. On May 22, 1992, the Convention on Biological Diversity was adopted at the Rio Earth Summit. Biodiversity is about more than plants, nonhuman animals, other organisms, and their ecosystems. It includes people. The goals put forth in the convention focus on conservation of biodiversity, sustainable use of resources, and fair and equitable sharing of benefits arising from the use of genetic resources. We all stand to reap social, environmental, and economic benefits from conservation of biodiversity. The United Nations General Assembly has declared May 22 as the International Day for Biological Diversity, to commemorate the day the convention was adopted.

MAY 23

Walk with a Turtle

Today is World Turtle Day, sponsored by American Tortoise Rescue since 2000. It's a day to celebrate turtles and increase awareness of their need for protection.

Folklore worldwide praises the turtle. The fact that turtles live a long time implies endurance, so we associate turtles with strength and stability. Longevity also suggests accumulated knowledge. We consider turtles to be wise and able to foresee the future. Their slow, deliberate movements suggest persistence and thoughtfulness. They live self-contained between their shells, conveying self-reliance. For cultures worldwide, turtles symbolize good fortune and protective power. Devotees wear turtle amulets, charms, and talismans to bring good luck, improve health, increase longevity, and ward off evil spirits or otherwise protect from harm. Turtles represent timelessness, serenity, and peace.

For a rewarding and unique way to celebrate World Turtle Day, consider signing on for a turtle conservation vacation. Options include protecting nesting sea turtles or helping hatchlings reach the ocean on Costa Rican beaches; caring for giant tortoises on the Galápagos Islands; transporting sick and injured gopher tortoises to rehabilitation centers in Florida; or rescuing injured desert tortoises in California. We all can celebrate the day by following the advice of American writer Bruce Feiler: "Take a walk with a turtle. And behold the world in pause." We would do well to emulate the turtle—tenacious yet tranquil.

MAY 24

Water or Rice Wine?

Each mid- to late-May, Cambodians celebrate an ancient harvest ritual called the Royal Ploughing Ceremony, to mark the beginning of rice growing season. Oxen are central to this event, because they are believed to influence the annual rice harvest. Traditionally, members of the royal family performed the ceremony by plowing the first furrows in the sacred rice field to appease the harvest gods and thus ensure fertility.

Today, the ceremony celebrates farming and encourages farmers to produce bountiful crops. The King generally appoints high-ranking officials to perform the ceremony. A man leads the yoke and wooden plow to which two sacred, highly adorned oxen have been harnessed. A woman follows behind and sows rice seeds in the furrow made by the oxen. After circling the rice field three times, they stop at a chapel, where Brahmins solicit protection from the

gods. The oxen are unharnessed and led to seven silver trays containing rice, corn, beans, sesame seeds, freshly cut grass, water, and rice wine. Royal soothsayers predict the bounty of the year's harvest depending on what the oxen eat and drink. If they choose rice, corn, beans, or sesame seeds, the harvest will be good. If they eat grass, cattle are likely to become sick. If the oxen drink water, rain will be plentiful and peace will reign in Cambodia. But if the oxen drink rice wine, droughts are likely and the kingdom will experience trouble. . . .

MAY 25

Alive and Well

By the turn of the twentieth century, wild birds in parts of the United States were threatened by illegal commercial hunting. Hunters poached game in one state and sold the meat in another state where hunting was legal. On May 25, 1900, President William McKinley signed the Lacey Act, the first federal law protecting wildlife. The act made it a crime to trade in animals that were illegally taken, possessed, transported, or sold. The Lacey Act has been amended several times, most recently in 2008 to include additional plants

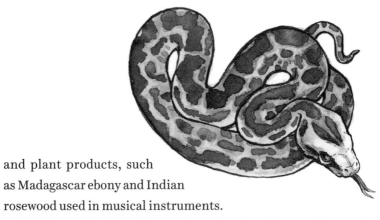

and plant products, such
as Madagascar ebony and Indian
rosewood used in musical instruments.

The Lacey Act also addresses introduction of nonnative species into the environment, because such species can become established and wreak havoc. Exotic animals might trample the habitat, compete with native species for food or space, introduce disease and parasites, or eat native species. One example is the Burmese python in the Florida Everglades. In 1979, one Burmese python was discovered in the Everglades, presumably released by a pet owner. By 2017, population estimates of Burmese pythons living in the Everglades range from 10,000 to 150,000. Some of their prey species, such as raccoons and rabbits, are declining. To prevent similar environmental disasters, in 2012 the Lacey Act listed four species of nonnative constrictor snakes as "injurious wildlife" and banned them from importation and interstate transportation: yellow anacondas, Burmese pythons, and northern and southern African pythons. The Lacey Act of 1900 is alive and well.

MAY 26

The Silent World

> The sea, once it casts its spell, holds one in its net of wonder
> forever.
> —JACQUES YVES COUSTEAU, French oceanographer and
>   filmmaker

Many of us look up into the sky with wonder. Others look down into the ocean and contemplate its origin, chemical and biological composition, and future. Jacques Cousteau was one of the latter. Cousteau learned to swim at the age of four, launching his lifelong fascination with oceans and aquatic life. Many people remember Cousteau as the person who introduced them to the ocean's wonders and allowed the sea to cast its spell. On May 26, 1956, Cousteau's documentary *The Silent World* premiered at the Cannes Film Festival. The film, based on Cousteau's best-selling book *The Silent World: A Story of Undersea Discovery and Adventure* (1953), was awarded the Palme d'Or ("Golden Palm"), the highest prize awarded each year at the festival. *The Silent World* was one of the first films to showcase the ocean depth in color through underwater cinematography. Tropical fish, whales, sharks, and coral reefs became more "real" for us.

Cousteau shared with us his love and wonder of the ocean through more than 120 television documentaries and 50 books. He founded the nonprofit Cousteau Society and its French counterpart, l'Équipe Cousteau, both of which continue to support research, education, and conservation of the world's now somewhat less mysterious oceans. Through his passion, Cousteau inspired us to respect the oceans and marine life.

MAY 27

A Little Off-key

On May 27, 1784, Wolfgang Amadeus Mozart bought a European starling from a pet shop in Vienna. In his expense diary, he recorded the transaction, a transcription of the 17-note melody the bird had sung, and the words *Das war schön* (That was beautiful!). No wonder Mozart was pleased. The melody was straight from the final movement of the Piano Concerto in G Major, K. 453, which he had just completed a few weeks earlier. The bird had learned the melody

from Mozart's concerto, but how?

Starlings have a rich vocal repertoire, and they repeat the sounds of other birds and human speech and weave them into their own soliloquies. Research by animal behaviorists Meredith West and Andrew King on starling vocal performance reveals that these birds quickly learn and mimic phrases and music. West and King suggest that Mozart may have visited the pet shop before May 27. Mozart hummed and whistled frequently, and he may have whistled the theme while in the store. West and King note that the starling could have sung the tune after a single hearing.

The performance was not flawless, however. The starling sang a G sharp where Mozart had written a G natural. West and King point out that this is typical starling behavior. The birds often improvise and sing off-key, rendering unique interpretations of what they hear. Mozart must have found that awfully humorous.

MAY 28

Whoopers

> The dancers raised their wings and lifted their feet, first one
> and then the other. They sunk their heads deep in their snowy

breasts, lifted them and sunk again.

—MARJORIE KINNAN RAWLINGS, *The Yearling*

Marjorie Rawlings' dancers are whooping cranes engaged in their elaborate mating dance. Today is Whooping Crane Day, a celebration of these 5-foot (1.5-m) tall, snowy white birds. The whooping crane story reads from tragedy to renewed hope. Before the arrival of European settlers, an estimated 15,000 whoopers danced across the US. By 1860, the number had plummeted to about 1400, caused by habitat loss and hunting. By 1941, only 15 cranes, comprising a single flock, migrated between northwestern Canada and the Gulf Coast of Texas. Conservation efforts paid off, and by 2005 the flock had increased to 214 cranes.

Biologists have facilitated a second migratory flock. Female whooping cranes usually lay two eggs, but a pair can care for only one chick. In 2001, biologists removed the second egg from nests, and, dressed in white crane costumes, raised the orphans. The chicks imprinted on the biologists and trailed after their foster parents. One day a faux adult crane drove an ultralight aircraft along the ground. The young birds ran behind. Later, the ultralight lifted into the air,

and the young whoopers followed. Eventually, the aircraft flew from Wisconsin to Florida, with the birds flying behind. Since 2001, every year a class of young whoopers have followed the aircraft from the Wisconsin breeding grounds to the Florida wintering grounds. The cranes fly back to Wisconsin on their own the following spring. As of August 2017, 95 cranes (not including the wild-hatched chicks of that year) make up the Florida-Wisconsin population, fostering hope for the species.

MAY 29

Time Is Relative

> When you are courting a nice girl, an hour seems like a second. When you sit on a red-hot cinder, a second seems like an hour. That's relativity.
> —ALBERT EINSTEIN

Dolania americana mayflies spend a year living as nymphs in sandy stream bottoms. Once they emerge, females have five minutes to find mates, copulate, and lay eggs in the stream. Then they die, having lived the shortest adult life span of any insect. Females are reproductive adults for less than 0.0009 percent of their lifetimes. In contrast, male *Dolania americana* mayflies spend a relatively long time as adults—six times as long as females, a whopping 30 minutes.

Worldwide, an average human female spends about 43 percent of her lifetime as a reproductive adult (about 32 of 74 years), of which she is fertile about 2112 days (5.5 days/month), or 7.8 percent of her life. Relative to the female mayfly, that's a lot, but relative to human males, it's a drop in the bucket. Human males have been known to father children into their 90s. It's all relative.

On May 29, 1919, a solar eclipse suddenly made Albert Einstein a worldwide celebrity. Sir Arthur Eddington, a British astronomer, measured the shift in position of stars during the eclipse to test Einstein's prediction that the gravity of the sun could bend the path of light. The prediction was confirmed, and Einstein's General Theory of Relativity was deemed correct. Scientists have continued to measure light-bending during eclipses, always with the same Einsteinian result. Space and time are relative.

MAY 30

"People Are Corn"

> When we plant corn we place seven or eight seeds in each hole. Of course, we don't need to grow that many plants for ourselves, but one plant is for the mouse and two are for the crow. They need to eat too, you know, and they like corn just as we do.
> —CLIFFORD BALENQUAH, Governor of Bakabi Hopi village
> (quote from *Expedition*, July 1991)

The Hopi from northeastern Arizona have dealt with vagaries of weather patterns in the semiarid climate of their high desert home for more than 2000 years. They believe that spirits called kachinas (spirit messengers) control nature, and that through prayer they can petition these supernatural beings to bring rain. Because the Hopi possess prayer feathers, corn pollen, and rituals they believe the kachinas desire, many traditional ceremonies involve mutual gift-giving.

During May, the Hopi plant their first crops—beans, pumpkins, and melons. By late May or early June, it is time to plant corn, a mainstay of their diet and a central focus of nearly every Hopi ceremony. Corn has spiritual significance for the Hopi, who view corn as a metaphor for life and say that "people are corn." The Hopi recognize that both people and corn begin as seeds, are nourished with sunlight, air, and water, grow and mature, and then die and return to the earth.

MAY 31

Ecotourism

> Study nature, love nature, stay close to nature. It will never
> fail you.
> —FRANK LLOYD WRIGHT, American architect

Surround yourself with the beauty of exotic rainforests, savannahs,
glaciers, mountains, prairies, oceans, and deserts. Count whales
and hatchling sea turtles for conservation. Photograph orchids,
hummingbirds, lemurs, and giraffes. Swim with penguins and sea
lions; snorkel among tropical fish, corals, sponges, and sea urchins.
Bird-watch from canopy walkways in Australia, Malaysia, South
Africa, and Costa Rica. Experience ecotourism—tourism based on
nature.

Ecotourism is an increasingly popular commercial venture. Ideally, it provides enriching experiences that promote greater appreciation for nature, empowers local communities to create sustainable development, and helps to conserve biodiversity. But sometimes ecotourism causes more harm than benefit. Animals are disturbed, vegetation is trampled, ornamental plants are stolen, and local people's economy and culture are forever changed. To prevent these negative consequences, in May 1990, The International Ecotourism Society (TIES) was founded with the goal of providing guidelines, training, technical assistance, and educational resources to encourage and support effective ecotourism. Organizations in more than 120 countries currently belong to TIES. In 2015, protected natural areas around the world collectively received an estimated 8 billion visits—more than one for every living person! Clearly, ecotourism must be practiced responsibly, to minimize our environmental impact.

JUNE 1

Carrier Pigeon Airmail

Humans have long appreciated the capability of carrier pigeons to return home when released in a new location, and thus their ability to relay messages. A frieze in the Egyptian temple of Medinat (ca. 1297 BCE) depicts a priest releasing four pigeons to send news of the pharaoh's coronation. Ancient Romans sent pigeons to relay the outcomes of chariot races. Genghis Khan used pigeons as messengers during his conquest of Eurasia in the late twelfth century. Pigeons delivered news to England of Napoleon's defeat during the Battle of Waterloo in 1815. The birds relayed messages during the Franco-Prussian war (1870) and both World Wars. During WWII, 26 pigeons received the Dickin Medal, the highest military decoration awarded by the United Kingdom for working animals during war.

Delivery by pigeon aerial post is not hampered by traffic jams

or blocked roads. Often a message is written on lightweight paper, rolled into a tube, and attached to the pigeon's leg. Pigeons transport laboratory specimens in tiny, unbreakable vials between hospitals in England, and between hospitals in France. They carry microfilm and memory sticks.

The carrier pigeon distance record was set by the Duke of Wellington's pigeon, which allegedly flew 6800 miles (11,000 km) from Namibia to London in 55 days. On June 1, 1845, the exhausted bird dropped from the sky, dead, one mile short of its destination. Today seems an appropriate day to honor these avian messengers for their stamina, dedication, and service.

JUNE 2

Animal Release

Today is part of the month-long Saka Dawa celebration in Tibet, a time to honor the Buddha. Saka Dawa occurs during the fourth Tibetan month (May-June in the Gregorian calendar). Good deeds performed during Saka Dawa are magnified millions of times over. One popular deed is buying and releasing animals destined to be killed or held captive, including fish, turtles, and birds from food markets and pet stores.

The rationale for animal release, *tshe thar* (*fang sheng* in Chinese), is the Buddha's teaching of compassion toward all living creatures. Beginning in the sixth century, Buddhists used public ceremonies as occasions to release animals destined for slaughter. Fish and turtles were liberated in temple ponds, and goats, sheep, cows, and horses were transferred to pastures. *Tshe thar* is believed to generate positive karma for the person performing the deed, bringing good fortune for this life and hope for the next.

Today, *tshe thar/fang sheng* raises ethical concerns. Wild birds are captured just so they can be released. Temple ponds are overcrowded with diseased turtles. Nonnative animals are released, posing a threat to native species. Understandably, conservationists, philosophers, and spiritual leaders have different perspectives on whether animal release is worth the environmental and ethical price of earned merit. A current movement in the United States encourages monks associated with Buddhist temples to work with certified wildlife rehabilitators. After animals are rehabilitated, Buddhists can bless and release the animals in an environmentally appropriate way.

JUNE 3

Earth, Ancient and Dynamic

During the latter part of the eighteenth century, popular belief in Europe held that Earth was created only 6000 years earlier. Human wickedness caused God to deliver a great flood. Fossils were the remains of animals that died during the flood, and Earth was repopulated by animals Noah gathered on the ark. Then in 1785 along came James Hutton, Scottish gentleman farmer/naturalist/ geologist, who argued that Earth is constantly being formed and re-formed through slow-moving volcanic processes, erosion, and sedimentation. These geological forces today are the same as those operating in the past. Only enormous amounts of time can explain the thicknesses of exposed rock layers. Heresy!

James Hutton, founder of modern geology, was born on June 3, 1726. Hutton's ideas have profoundly affected geological and biological thought, but for several decades after his writings many scientists and philosophers strongly disagreed with him. Instead, they believed in catastrophism, the idea that geological changes were caused by natural disasters such as volcanic eruptions and floods. During the 1830s, geologist Charles Lyell further developed Hutton's ideas (now known as the principle of uniformitarianism) in his 3-volume book *Principles of Geology* (1830–1833). Hutton's concept of a dynamic and ancient Earth many millions of years old provided the eons of time Charles Darwin needed to explain what he saw in the fossil record.

JUNE 4

King of Cheese

According to legend, one day a shepherd boy in southern France was eating bread and sheep's milk cheese when he saw a beautiful girl. He left his lunch in a nearby cave and ran to meet her. Months later, the lad returned to the cave and found his cheese riddled with blue mold. After tasting it and finding the tangy, rich flavor delicious, he returned to his village shouting "A miracle, a miracle!" The villagers were equally impressed and soon began storing their cheese in the caves. These same caves of Roquefort-sur-Soulzon are still used today to make Roquefort cheese. The blue veins in Roquefort are the fungus *Penicillium roquefortii*, found naturally in the cave soil.

On June 4, 1411, King Charles VI of France granted sole rights to the village of Roquefort-sur-Soulzon to age Roquefort blue cheese. Now, more than six centuries later, 3 million wheels (nearly 19,000 tons) of Roquefort are produced each year—every one aged in the Mont Combalou caves of Roquefort-sur-Soulzon. Roquefort is

made from the milk of Lacaune, Manech, and Basco-Béarnaise breeds of sheep, and the *Penicillium roquefortii* fungus must come from the natural caves of Roquefort-sur-Soulzon.

For centuries, residents of southern France applied Roquefort cheese to wounds to prevent infection. They knew the cheese was effective, even if people elsewhere viewed the practice as quackery. Once penicillin was discovered in 1928, the medicinal benefits of the cheese made sense. More recently, a 2012 study revealed that Roquefort also contains anti-inflammatory compounds. *Penicillium roquefortii* is one good fungus among us!

JUNE 5

World Environment Day

> Did you ever stop to notice
> The crying Earth, the weeping shores?
> —MICHAEL JACKSON, "Earth Song"

Every June 5, the United Nations Environment Programme sponsors World Environment Day to encourage us to protect Earth. Designated in 1972, World Environment Day is now celebrated in over 100 countries. Each year has a different theme and host country. Mexico hosted the event in 2009, with the theme "Your Planet Needs You—Unite to Combat Climate Change." That year, Michael Jackson's "Earth Song" was declared the World Environment Day song.

Other examples of themes and host countries include "Development without Destruction" (Bangladesh); "Poverty and the Environment—Breaking the Vicious Circle" (China); "For Life on Earth—Save Our Seas" (Russia); "Green Cities—Plan for the Planet" (United States); "Forests—Nature at Your Service" (India); "Raise Your Voice, Not the Sea Level" (Barbados); and "Prohibit Illegal Trade of Threatened Wildlife" (Angola). In 2017, World Environment Day was held in Canada, where the theme was "Connecting People to Nature."

World Environment Day is an invitation to think about how you can help protect Earth. Begin simply. Resolve to double your efforts to compost, reuse and recycle, and conserve water and electricity. Walk or bike more; drive less. Get involved with cleanup; help to educate the public. The health of the planet depends on each of us.

JUNE 6

Red, Red Rose

> Oh my Luve is like a red, red rose
> That's newly sprung in June.
> —ROBERT BURNS, "A Red, Red Rose"

Throughout history and worldwide, people have adored roses. Funerary wreaths of roses have been found in ancient Egyptian tombs. Admirers showered returning victorious Roman armies with rose petals, and Cleopatra filled her bedroom chamber with rose petals to seduce Marc Antony.

Diverse cultures link roses with blood. According to Greek mythology, the rose originated from the comingling of the tears of Aphrodite, goddess of love, and the blood of her lover, Adonis, after he was wounded by a wild boar. For the ancient Romans, the rose was a sign for Venus, goddess of love. Legend tells that one day her son, Cupid, shot arrows into the rose garden and caused the roses to grow thorns. Venus walked through the garden and pricked her foot on a thorn. Her blood turned the roses red. An Arabic legend tells that originally all roses were white. One day a nightingale fell in love with a perfect white rose. He pressed himself against the flower, and the thorns pierced his heart. His blood turned roses red forever.

The rose is the birth flower for June, symbolic of appreciation and love. Red roses are associated with romance and passion; white with innocence, humility, and remembrance; pink with admiration and sympathy; and yellow with joy and friendship.

JUNE 7

Spring Orgy and Feast

During May and June high tides of full and new moons, over a million horseshoe crabs invade the beaches of Delaware Bay on the US Atlantic coast. Males lie in wait in shallow water and clasp incoming females, and the pairs crawl ashore. Each female digs a hole, deposits several thousand small, greenish eggs, and drags her sperm donor over them. Waves wash sand over the nest and cover the eggs. The pair moves on and repeats the process. Each female lays 80,000 or more eggs during the season. But just because a male has claimed a female, it doesn't mean that he will fertilize all her eggs. This is a frenetic orgy during which four, five, or more males might simultaneously clasp one female.

Waves disturb nests and expose the eggs. Female horseshoe crabs clawing in the sand disrupt other nests and fling eggs to the surface. Each egg is a packet of fat—perfect food for an animal needing to replenish its fat reserves. Coinciding with the orgy, hundreds of thousands of shorebirds migrating from South America to Canada stop off to feast on horseshoe crab eggs. Red knots, ruddy turnstones, and other birds consume about 539 metric tons of the eggs during their visit, the product of an estimated 1,820,000 horseshoe crabs. This combined spectacle of horseshoe crabs and shorebirds supports a multimillion-dollar ecotourism industry for Delaware Bay. The dramatic orgy and feast illustrate two primal instincts—reproduction and eating—in a hedonistic show of grandeur.

JUNE 8

An Ocean Celebration

> With every drop of water you drink, every breath you take,
> you're connected to the sea. No matter where on Earth you live.
> —SYLVIA EARLE, marine biologist

June 8 is World Oceans Day—a global celebration of the World Ocean, composed of the Pacific, Atlantic, Indian, Southern, and Artic Oceans. Today is a day to appreciate the ocean's allure and its role in our lives, and to highlight conservation of the ocean's resources.

For many of us, the ocean conveys a sense of serenity and constancy. Underneath the ocean's surface, however, a rich diversity of life struggles to survive hungry predators. Life there is anything but serene. Predator defenses are as varied as the organisms themselves. A hagfish deters predators by oozing slime from hundreds of glands along its body—enough to fill seven buckets within minutes. When a sea cucumber senses danger, it expels its insides—sticky spaghetti that can entangle its attacker. (The sea cucumber later regenerates its internal organs.) A cone snail, one of the most venomous of all animals, pierces would-be predators with a harpoon-like "tooth," causing pain, paralysis, and death. Jellyfish tentacles are covered with venom-filled stinging cells (nematocysts) that fire when touched. An octopus's response to a predator ranges from dramatically changing color to ejecting ink as a smokescreen. Imagine the amazing animal defenses we have yet to discover—another reason to protect the oceans' resources.

JUNE 9

Miracle of the Gulls

Brigham Young led the first group of nearly 2000 Latter-day Saints (Mormons) into Salt Lake Valley, Utah, in July 1847. During that summer and fall the settlers planted winter wheat. Another 2400 Mormons arrived in 1848. That spring the settlers planted corn, beans, melons, squash, and other crops. According to Mormon lore, by May 22 hordes of flightless katydids (later dubbed "Mormon crickets") invaded their fields and devoured the crops. In early June, seagulls arrived in large flocks and feasted on the katydids. On June 9, 1848, a letter sent to Brigham Young described the scenario: "The sea gulls have come in large flocks from the lake and sweep the crickets as they go; it seems the hand of the Lord is in our favor." Reportedly, the gulls gorged on katydids, drank water, regurgitated, and gorged on more katydids. The gulls rescued the first harvest for the new settlers to the Salt Lake Valley.

Sculptor Mahonri M. Young, grandson of Brigham Young, cast a bronze statue of two seagulls in honor of the birds that saved the Mormons' first crops. This statue, called Seagull Monument, was erected in front of the Salt Lake Assembly Hall on Temple Square in Salt Lake City and dedicated in October 1913. In 1955, the California gull was declared the state bird of Utah to further honor the bird that facilitated survival of Utah's pioneer settlers. Large colonies of California gulls still nest around Great Salt Lake.

JUNE 10

Biophilia

How we enjoy a day with no obligations! No job demands, no appointments, no home-improvement chores. Many of us choose to spend these days outside. We photograph flowers or wildlife, hike, mountain bike, hunt, fish, or canoe, reinforcing our bond with nature. The concept of "love of living things," or *biophilia*, goes back at least as far as Aristotle, more than 2000 years ago. It wasn't, however, something that most of us thought much about. That is, not until Edward Osborne Wilson, born on June 10, 1929, renowned naturalist and Harvard professor, crusader of preserving biodiversity, Pulitzer Prize–winning author of over a dozen books, and the world's expert on ants, published his 1984 book *Biophilia*.

In his book, Wilson hypothesized that humans have an innate tendency to affiliate with nature. He suggested that humans subconsciously seek connections with the rest of life, that we evolved with this trait, and that it is still ingrained in our genome. In subsequent years, Wilson revised his concept of biophilia, suggesting that our urge to connect with the natural world is also learned.

The idea that biophilia may be partly a learned state of mind increases the urgency to cultivate a love of nature in our children. The more our children embrace the natural world, the more they will protect it.

JUNE 11

To Clone a Dinosaur

The idea of restoring dinosaurs to Earth has long fascinated writers, scientists, and the public. Steven Spielberg's movie *Jurassic Park*, based on Michael Crichton's book of the same title, opened on June 11, 1993. The film grossed more than $900 million worldwide during its original theatrical run, the highest-grossing movie to that time. A billionaire and his team of geneticists have created a theme park of cloned dinosaurs on an islet near Costa Rica. The scientists extracted dinosaur DNA from mosquitoes preserved in amber and filled in the missing dinosaur genomes with frog DNA to resurrect *Velociraptor*, *Triceratops*, *Tyrannosaurus rex*, and more.

Perhaps disappointingly, real-life scientists have never recovered dinosaur DNA, and without that we cannot clone a dinosaur. Mary Schweitzer, a molecular paleontologist at North Carolina State University, found DNA in dinosaur bone, but the identity remains a mystery because the DNA could not be recovered and sequenced. It could belong to microbes, or to paleontologists who handled the fossils. An animal's DNA begins to decay the moment the animal dies. Research suggests that DNA has a half-life of about 520 years; thus, a dinosaur's DNA would have been completely destroyed within 7 million years of the animal's death—far short of

the 65 million years since the end of the dinosaurs. Schweitzer notes that many problems would have to be overcome to clone a dinosaur. Getting the DNA would be the easy part. At least we have *Jurassic Park* to satisfy our fantasy.

JUNE 12

The Egyptian Vulture: Doting and Clever

Ancient Egyptians honored Mut, the mother goddess of Thebes, each year on June 12. The Festival of Mut was one of the most popular holidays during the New Kingdom (1570–1070 BCE). As part of the festivities, a statue of Mut was placed on a boat and sailed around the sacred lake at her temple. Mut was depicted either as a vulture or as a woman with vulture wings. Ancient Egyptians viewed the vulture as the most fiercely protective, nourishing, and loving of all mothers, a belief that suggests Egyptians were impressed with the maternal behavior they observed from their local vultures. In time, Mut's name became the ancient Egyptian word for mother,

mwt. When Thebes became the capital of Egypt, Mut was seen as the mother of the nation and was worshipped until Rome's conquest of Egypt in 30 BCE.

Egyptian vultures generally are monogamous; they nest on cliffs and rocky slopes, constructing nests that measure 5 feet (1.5 m) across. Both parents incubate their two eggs for about 40 days, and they share in feeding the chicks until they fledge at about 70 to 85 days old. Although they eat mainly carrion, Egyptian vultures also eat other birds' eggs, including thick-shelled ostrich eggs. In 1966, Jane Goodall and several others observed Egyptian vultures throwing stones at ostrich eggs to break them open—their own type of "tool." One wonders if the ancient Egyptians knew of the doting parents' cleverness!

JUNE 13

Wild Swans

> But now they drift on the still water,
> Mysterious, beautiful;
> Among what rushes will they build,
> By what lake's edge or pool
> Delight men's eyes when I awake some day
> To find they have flown away?
> —WILLIAM BUTLER YEATS, "The Wild Swans at Coole"

William Butler Yeats was born on June 13, 1865, in Dublin. A proud Irishman, Yeats based much of his poetry on Irish legends, folklore, and ballads and was inspired by the natural beauty of his country. Scholars have hailed Yeats as one of the greatest poets of the twentieth century. He also wrote essays, plays, short stories, and novels, and was awarded the Nobel Prize for Literature in 1923, the first Irishman so honored. Ireland declared 2015 The Year of Yeats, marking the 150th anniversary of his birth.

At the age of 51, Yeats wrote one of his best-loved poems, "The Wild Swans at Coole," while visiting the estate of Coole Park, County Galway. Yeats is melancholy, aware of his advancing age, reeling from romantic rejection, and mourning the political chaos of his country being torn apart by rebellion against the British. He reflects that all has changed since his last visit to Coole, 19 years earlier, yet the swans' hearts have not grown old like his. The swans will return to the lake each autumn, as they always have. Yeats' poem offers reassurance for nature's continuity, even in the midst of our own changes.

JUNE 14

Our Magical Rivers

> No man ever steps in the same river twice, for it's not the same river and he's not the same man.
> —HERACLITUS (paraphrased quote from the ancient Greek philosopher)

Folklore imbues rivers with magical properties. The rushing water can heal, cleanse, and protect against evil. We can escape demons and witches by swimming against a river's current. Stones polished smooth from tumbling cure our ailments. Rivers are associated with rebirth and rejuvenation because they change constantly. They symbolize independence and self-assurance: they always overcome obstacles as they head toward their destination. Rivers represent the flow of time and progression of life: they are born in high mountain springs or from melting snow, and they die at the ocean.

The United States boasts nearly 3 million miles (more than 4.8 million km) of running water. These waterways range from nameless crystal-clear streams meandering through columbine, little pink elephant, paintbrush, and lupine above timberline, to muddy floodwaters sweeping down desert arroyos, to the mighty Mississippi River and the grand Colorado River. June is National Rivers Month in the US, with the three-fold goal of getting people to learn about, celebrate, and clean up our waterways. It's a great time to perfect your fly-fishing skills; raft through some white water, kayak, or canoe; or camp, hike, or birdwatch alongside your favorite river. Enjoy the river's sense of timelessness, its music, its energy. Follow the river as it weaves in and out of the landscape, leading you around the next bend.

JUNE 15

Flashing Rear Ends

Flashing their rear ends in search of mates, fireflies (also known as lightning bugs) add beauty and a touch of magic to summer nighttime. The epithets fireflies and lightning bugs are both misnomers. They are neither flies nor bugs, but rather soft-bodied beetles. More than 2000 species of fireflies are distributed on every continent except Antarctica.

There is much to admire about fireflies. They don't bite or sting. They don't disturb our sleep with raucous song. And their larvae eat slugs and other garden pests. According to legend, Chinese scholars too poor to afford lamp oil studied by the light given off by captive fireflies. Still today indigenous peoples in South America tie plastic bags filled with fireflies to their ankles, providing living flashlights for navigating the nighttime jungle.

The Japanese have long celebrated the *hotaru* (fireflies), considered to be souls of deceased samurai. They are also symbolic of silent but passionate love or of fleeting and unrequited love. Every June, firefly festivals held throughout Japan encourage people to gather and admire the twinkling *hotaru*. Larvae that have been living underground have pupated. In June the pupae split, and adult fireflies

emerge. They will live only a short two weeks or so, but what beauty and magical moments they will share with us during their search for mates!

The Green Turtle Story

Today is World Sea Turtle Day, a day to honor these extraordinary reptiles. The celebration occurs on the birthday of Archie Carr, father of sea turtle biology and a champion of their conservation, born in 1909. Archie spent much of his life studying and telling the story of green turtles.

On a moonlit night, a green turtle rises with a wave. She hauls her 330-pound (150-kg) body out of the ocean and crawls to sand

above the high tide line. Using her front flippers, she digs a basin to rest in while she lays her eggs. With her hind flippers, she digs a nest hole in the basin. She lays 115 eggs, each the size and color of a ping-pong ball, and then fills the egg chamber with sand. Slowly, she crawls back to the ocean.

Sixty days later, a few baby turtles break out from their leathery eggshells. Once most have hatched, the babies thrash about. Their frantic activity causes the sandy ceiling to collapse. That night, the hatchlings bubble out of the sand into darkness. They instinctively head toward the brightest horizon—moonlight and starlight reflecting off the ocean surface. Coyotes, dogs, raccoons, gulls, night herons, and ghost crabs eagerly await turtle dinners. Hatchlings that make it into the ocean encounter sharks and predatory fish. This cycle of events—egg-laying to baby turtles racing to the ocean—has been happening for over 100 million years.

JUNE 17

The "African Unicorn"

For centuries, folk tales of a horned horse—a strange chimera part zebra, part donkey, part giraffe—filtered out from central Africa to the Western world. For people living in the Ituri Forest of the Democratic Republic of the Congo (DRC), this animal was powerful and mysterious. But when European expeditions could find no such animal, most Western biologists discounted the tales and rejected the animal's existence. It was deemed mythical and dubbed the "African unicorn." One naturalist, however, suggested that if it did exist, it was *Hipparion*, a genus of horse that lived millions of years ago.

Then, in 1901 Sir Harry Johnston, explorer and colonial administrator, began a search for the African unicorn in the Ituri Forest. He and his party were successful. Johnston sent the skin and two skulls of the mystery animal to the British Museum. They arrived on June 17, 1901, and the following day Professor E. Ray Lankester described the specimens as a new genus and species: *Okapia johnstoni*.

The okapi, an herbivore related to the giraffe, is one of the last large mammals to be discovered from Africa. Congolese from the

DRC are justifiably proud of their endemic okapi, and they chose it as their national animal. The story of how the Western world came to believe in the "African unicorn" is a valuable lesson in how much we can learn from local folklore.

Animal Coins

Animals have appeared on coins since the beginning of coinage about 700 BCE. Ancient Greek and Roman coins depicted octopuses, crabs, fishes, snakes, doves, owls, eagles, dolphins, deer, pigs, wolves, rams, horses, and lions. Some of today's most popular animal coins include the Australian Gold Nugget featuring a kangaroo; the Chinese gold and silver panda coins; the Mexican peso with its coat of arms of an eagle eating a snake; the South African Gold Krugerrand with a springbok antelope; and the US gold and silver bald eagle coins and buffalo nickel. On June 18, 1987, the Australian government approved the minting of the platinum koala coin, one of the few platinum coins still issued.

Although birds and mammals are the most common vertebrates on present-day coins, reptiles are popular as well. Sea turtles appear on coins from Ascension, Brazil, Cape Verde, Cayman Islands, Colombia, Cook Islands, Fiji, Maldives, and Turkey. Crocodiles are featured on coins from Australia, Cuba, Gambia, Jamaica, and Papua New Guinea. Snakes appear on coins from Australia, North Korea, Somalia, and Russia. The tuatara graces a New Zealand coin, and lizards are depicted on coins from Australia, Indonesia, Fiji, Poland, and Russia.

Animals have a huge impact on our lives, from keeping us company, to working for us, to providing protein. Religions, language, folklore, visual art, music, and literature reflect our bond with other animals. It's no wonder that we also incorporate nonhuman animals into objects we use every day, such as coins.

JUNE 19

Burped Up by Dad

In many countries, the third Sunday in June is Father's Day. It's a good day to recognize the nonhuman fathers that also play critical roles in their offspring's lives. Darwin's frog is one of nature's extraordinary fathers. This frog, found only in Argentina and Chile, was shrouded in mystery for decades after its discovery.

In 1834, Charles Darwin found an unusual little frog with a Pinocchio-like snout in southern Chile. Unable to identify the frog, he sent specimens to England. In 1841, French scientists André Marie Constant Duméril and Gabriel Bibron named the new species *Rhinoderma darwinii*. A few years later French zoologist Antoine Guichenot examined Darwin's specimens and discovered tadpoles inside one of the frogs. He assumed the frog was a female and that Darwin's frogs are viviparous. Guichenot was mistaken. In 1872, Spanish zoologist Jiménez de la Espada examined additional specimens and discovered tadpoles in the vocal sacs of five males!

For the next 60 years, how the young got into the vocal sac remained a mystery. Then, in the 1930s, Chilean scientist Ottmar Wilhelm placed some Darwin's frogs in a terrarium and watched. One day the female laid eggs. Several weeks later, the embryos turned somersaults inside their capsules. Just before they would have hatched,

Dad slurped the eggs into his mouth. Instead of sliding into Dad's stomach via the esophagus, the eggs slid through his vocal slits and into his vocal sac. And there they hatched. We now know that Dad broods his tadpoles for about two months. When the time is right, he opens his mouth and miniature froglets hop out.

JUNE 20

In Honor of the Eagle

> Houston, Tranquility Base here. The Eagle has landed.
> —NEIL ARMSTRONG, American astronaut

On June 20, 1782, the Continental Congress adopted the design of the Great Seal of the United States of America: a bald eagle grasping a bundle of 13 arrows in its left talon and an olive branch in its right talon. More than 170 years later, President John F. Kennedy declared, "The Founding Fathers made an appropriate choice when they selected the bald eagle as the emblem of the nation. The fierce beauty and proud independence of this great bird aptly symbolizes the strength and freedom of America." Appropriately, the lunar module that allowed Neil Armstrong and Buzz Aldrin to land on the moon in July 1969 was named "Eagle."

Native American cultures have revered the bald eagle for much longer. For many First Peoples, bald eagles (along with golden eagles) symbolize wisdom and strength, and are believed to be medicine birds with special healing powers. Eagle feathers are considered sacred and thus play central roles in traditional ceremonies. Bald eagles soar high and disappear behind the clouds, so many tribes view these birds as spiritual messengers from the people to the gods and imbue them with powers to control the rain. Cherokee, Comanche, Hopi, Choctaw, and many other tribes perform an Eagle Dance—a request for eagles to carry prayers for rain or peace to the gods.

Today is National Eagle Day, a day for diverse cultures in the United States to honor the bald eagle.

JUNE 21

Power of the Sun

Summer solstice—the time when Earth's rotational axis is most inclined toward the sun—occurs in the Northern Hemisphere on June 20, 21, or 22. Also known as Midsummer, it is the longest day of the year (except in polar regions where daylight is continuous for days to months around the summer solstice).

People have long celebrated the summer solstice, associated with Earth's rebirth. For ancient Egyptians, the summer solstice announced that Sirius, the brightest star in the night sky, would soon cause the Nile to flood and nourish the land. Ancient Chinese celebrated Earth, femininity, and the upcoming force of *yin* on this day. Ancient Greeks honored their agricultural god Cronus on the summer solstice, and Romans honored Vesta, goddess of the hearth and protectress of marriage and virginity. With their crops newly planted, Vikings and other Nordic seafarers conducted foreign trade, met to discuss legal matters and resolve disputes, and sailed off for fishing and raiding expeditions around the summer solstice; it was a time for activity while the power of the sun was greatest. Cultures worldwide still welcome Midsummer with rituals, bonfires, feasts, and festivals to celebrate nature, the power of the sun, the changing seasons, and the cycle of life, death, and new beginnings.

JUNE 22

Nalukataq

Every spring and fall, bowhead whales migrate along the coast of northern Alaska. The area's Iñupiat whaling communities have a special concession that allows them to hunt a total of 50 bowheads each year. Before you cry "Foul!," realize that these indigenous peoples have been hunting bowheads sustainably for more than 1000 years. Following a successful whale-hunting season, the Iñupiat celebrate a mid-to-late June harvest festival they call Nalukataq. The event gives thanks to these animals that sustain the people and their culture, celebrates the bounty of fresh meat, and provides an opportunity for whaling captains to distribute whale meat and blubber to community members.

The festival begins with a prayer and raising of the ships' flags, followed by bread, coffee, and goose and caribou soups. Participants sing and tell stories, and eventually the whaling crews distribute meat, beginning with frozen raw meat cut into cubes and thin strips of flukes (the two lobes of a whale's tail). Following a feast of whale meat and fried blubber, the Nalukataq blanket toss begins. A length of rope extending from each corner of a blanket sewn from several bearded seal skins is pulled tightly between four wooden beams to raise the blanket to about waist height. People encircle the blanket and pull out on it to throw blanket dancers—first the captains and their wives—into the air. While airborne, the dancers throw candy to children waiting below. The festival ends with a traditional dance, songs, and a final prayer.

JUNE 23

The Power and Potency of Herbs

John the Baptist is believed to have been born six months before Jesus, which makes his designated birthday June 24. St. John's Eve, June 23, is celebrated in many countries with feasting, dancing, traditional songs, and bonfires.

Europeans have long collected medicinal plants on St. John's Eve, a time when the power and potency of herbs were thought to be greatest. Throughout Europe, St. John's wort was picked on St. John's Eve and hung in windows to expel demons, ward off witches, and protect against sickness and harm. Mugwort was collected on the morning of St. John's Eve and smoked over bonfires to strengthen the herb's powers. Garlands made from the strengthened mugwort were hung over windows and doors to ward off evil forces. During St. John's Eve celebrations, Germans wore garlands of mugwort to attract good fortune. The Irish gathered mallow leaves and stems on St. John's Eve. They touched every person they encountered with the herb to protect against sickness and evil influences during the coming year, and then they burned the herbs in bonfires. On St. John's Eve, the French passed springs of mullein across bonfire flames to protect cattle from sickness and sorcery.

We each try to control our own destiny—fortune, happiness, health, and death—through shamanic intervention, prayer, offerings, professional medical care, and/or folk medicine. Medicinal herbs have long been part of the mix.

JUNE 24

Gone, but Not Forgotten

Humans have seriously reduced populations of Galápagos tortoises. During the seventeenth century, buccaneers and whalers collected an estimated 100,000 tortoises from the Galápagos Islands for food to be eaten during their long voyages home. Animals accidentally and intentionally introduced to the islands have wreaked havoc. Rats eat tortoise eggs and hatchlings, and dogs and cats eat hatchlings. Pigs destroy the habitat, and goats compete with tortoises for food. By November 1971, an estimated 40,000 feral goats had devastated the vegetation on Pinta Island when a biologist found a lone Pinta Island tortoise. Nicknamed George, the tortoise was relocated to the Charles Darwin Research Station on Santa Cruz Island in hopes of finding him a mate. No other Pinta Island tortoise was ever found, wild or captive.

Lonesome George, believed to be the last Pinta Island tortoise, was found slumped in his corral on June 24, 2012. George died of old age, at more than 100 years. The President of Ecuador, Rafael Correa, mourned his death in a public address to the nation and expressed the hope that someday Lonesome George might be cloned. Soon after George died, scientists collected tissues for that eventual possibility. His frozen body was shipped to the American Museum of Natural History in New York City, where taxidermists preserved him. For nearly three months during 2014–2015, thousands of vis-

itors to the museum paid their respects to the world's most famous tortoise. Lonesome George will not be forgotten. He now resides back home on Santa Cruz Island where hundreds of thousands of visitors will pay tribute to him and meditate on the tragedy of human-caused extinction.

JUNE 25

National Catfish Day

> Farm-raised catfish have come a long way from their bottom-feeding ancestors.
> —PRESIDENT RONALD REAGAN, Proclamation 5672

In 1987, President Ronald Reagan declared June 25 as National Catfish Day, to celebrate the value of US farm-raised catfish (mostly channel catfish and blue catfish). Many fish tend to taste like the food they eat, so it's no surprise that bottom feeders like channel and blue catfish can taste "muddy." In the wild, these catfish are both scavengers and opportunistic predators on other fish and invertebrates like worms and crayfish. Farm-raised channel and blue catfish are fed scientifically-formulated, high-protein pellets that float on the water surface. This diet gives the fish a consistently mild, slightly sweet flavor.

President Reagan ended the Presidential Proclamation by stating, "I call upon the people of the United States to observe this day with appropriate ceremonies and activities." Accordingly, today would be a fine time to invite friends for a backyard catfish fry, complete with hushpuppies and tartar sauce. Or, perhaps an ethnic dish: grilled catfish fajitas with chipotle salsa, or jalapeño and honey BBQ catfish; catfish bisque, or catfish en papillote; catfish Alfredo, or catfish Tuscany; Cajun pecan-crusted catfish, or catfish etouffee; or spicy catfish strips with Thai peanut dipping sauce. Eat heartily, with the satisfaction of knowing that US farm-raised catfish are sustainable and environmentally friendly.

JUNE 26

Camping in Style

> Glamping: A form of camping involving accommodations and facilities more luxurious than those associated with traditional camping.
> —*Oxford English Dictionary*

The *Oxford English Dictionary* (OED) is updated quarterly. In June 2016, one of the 1000 new words added was *glamping*, used informally since about 2005. The OED gives the derivation of glamping as "Early 21st century; blend of glamorous and camping."

Have you *never* enjoyed sleeping on damp, lumpy ground? Schlepping a week's worth of granola, energy bars, and trail mix; instant eggs, rice, and potatoes; freeze-dried lasagna and vegetarian chili? Being awakened by water dripping through a leaking tent? Did you used to love primitive camping, but now in your eighth decade find it hard to squat on the ground and even harder to get back up?

Take heart. You can experience nature—hike, photograph, paint, fish, canoe, swim—and return to your luxurious canvas accommodation provided by a commercial enterprise. Sleep in a king-size bed, relax in a leather chair and read by interior ceiling lights, order a chef- prepared meal delivered to your tent, and enjoy an en suite bathroom with hot water. Worldwide, prices for glamping accommodations range from less than $100/night for the tent to over $3000/night per person, depending on the amenities and location. Purists might scoff at glamping, but the experience still allows us to connect with nature.

JUNE 27

An Inordinate Fondness for Beetles

Every other year, the UK celebrates National Insect Week during the last week in June. Sponsored by the Royal Entomological Society, the aim is to educate the public about the value of insects and their need for conservation. Local activities might include an insect photography contest, butterfly walks, and talks by entomologists. Beetles are often highlighted because they are abundant and many are spectacular. Beetles make up 40 percent of all species of insects— about 450,000 species. That's nearly 25 percent of all known animal species! When asked if there was anything one could conclude about God from studying natural history, the British evolutionary biologist and geneticist J. B. S. Haldane replied that the Creator had "an inordinate fondness for beetles."

One fascinating group of beetles often featured during National Insect Week is dung beetles (family Scarabaeidae). These beetles feed partially or exclusively on dung and lay their eggs in it, providing the larvae with a ready source of food. The sacred scarab (*Scarabaeus sacer*) from Egypt is a dung ball roller. When a male finds a fresh cow pie, he plunges in and forms a sphere. He rolls it until he finds a suitable resting spot, where he buries it. And there his mate will lay her eggs. Ancient Egyptians sometimes represented their sun god, Ra, as a great scarab rolling the sun like a dung ball

across the heavens. Egyptians often placed these scarabs, symbol of resurrection and immortality, in tombs with their deceased. Such was the value of a beetle.

JUNE 28

Listen to the Frogs

Do you enjoy listening to calling frogs? If so, FrogWatch USA might be for you. Founded in 1998, this citizen science project is an opportunity to learn about your local frogs and wetlands, while contributing data to help conserve frogs and their habitats. June, peak calling time for many frog species in the Northern Hemisphere, is a great time to participate in FrogWatch.

The protocol is simple. After you select and register a site to monitor, you are trained to identify the calls of frogs in the area. You visit your site sometime between 30 minutes after sunset and 1:00 a.m. and remain quiet for at least two minutes after you arrive. During the next three minutes you cup your hands around your ears and listen. You record the species calling and the intensity with which you heard the calls. You do this as often as you can during peak frog calling activity in your area.

This standardized protocol allows data to be compared across time and geographic location. It's a win-win for everyone: fun for participants, valuable for scientists, and a conservation boon for frogs. There may be a FrogWatch chapter near you. By 2017, there were 145 chapters in 41 states and the District of Columbia. If you don't have a chapter nearby, why not contact your local zoo, aquarium, nature center, or conservation organization and encourage them to host one. For Canadians, there's FrogWatch Canada.

JUNE 29

Strawberry Fields

Look at a strawberry and what springs to mind? Many see a heart. Perhaps that image is what led to the widespread belief in the medicinal value of strawberries. As long ago as 2600 BCE, the Chinese steeped strawberry leaves to detoxify the body and slow the effects of aging. Romans ate strawberries to lift the spirits and treat digestive ailments. Roman mythology linked strawberries with Venus, advocating that strawberries shared with another person will induce love. People worldwide have believed that strawberries increase fertility.

In the United States, strawberry festivals held from California to New York give thanks for a plentiful harvest. Cedarburg, Wisconsin, holds its strawberry festival every late June. The town celebrated its 32nd strawberry fest in 2017. Each year up to 100,000 fans come to enjoy the all-you-can-eat strawberry pancake breakfast, strawberry slushies and smoothies, strawberry crepes—everything imaginable, from strawberry wine to strawberry bratwurst. For the competitive, there are strawberry shortcake–eating and strawberry bubblegum–blowing contests.

Strawberries may not increase fertility, but these berries packed with vitamin C, fiber, and antioxidants do lift the spirits. Enjoy chocolate-covered strawberries knowing that the heart-shaped fruits lower blood pressure and increase HDL (the good cholesterol). Try eating these berries in the British tradition of strawberries and cream, or as the Greeks do—strawberries sprinkled with sugar and dipped in brandy.

JUNE 30

Sea Horses

The last weekend of June sees the two-day Shrimp Festival in Oostduinkerke, on the Belgian coast of West Flanders. Oostduinkerke is a protected natural reserve, surrounded by 240 acres of dunes. It is also the only place in the world where shrimpers currently gather their catch from horseback. Men decked out in bright yellow slickers and overalls ride their strong Belgian draft horses into the ocean about two hours before low tide. Each horse, breast-deep in the surf, drags a shrimp-catching net held open by two wooden planks.

Five hundred years ago, coastal farmers practiced horseback fishing along the coast of the North Sea, in the Netherlands, Britain, and France, to catch fish for fertilizing their fields. Those farms have long since been replaced by seaside property development, but the shrimpers in Oostduinkerke continue the tradition. They catch shrimp for human consumption—not as a lucrative livelihood (the price of their shrimp does not turn a profit), but to keep their shrimping/equestrian tradition alive.

Some men call the relationship with their Belgian draft horse a love story. Many horses are nervous about the strange sights, sounds, and smells of the ocean and the waves crashing against their legs and flanks. They take one look at the waves and gallop back to shore. Once a bold, ocean-loving horse is found, a shrimper values his horse for life.

JULY 1

Zoos, Education, and Conservation

On July 1, 1874, the first US zoo opened in Philadelphia, Pennsylvania. During the first year of operation, the zoo housed 813 animals and hosted over 228,000 visitors. Admission was $0.25 for adults and $0.10 for children.

Zoos began primarily as places to exhibit exotic animals for our enjoyment. They still do this, but they now offer more. We learn about the animals' native habitats, natural history, and conservation status. Polls find that after a zoo visit, many visitors reflect on how they, personally, can help reverse the loss of biodiversity. Zoos contribute to wildlife conservation and have saved some species from extinction through captive breeding programs. For example, black-footed ferrets have made a comeback thanks to captive breeding efforts at the Phoenix Zoo and other facilities. The scimitar-horned oryx, at one time widespread across northern Africa, was declared extinct in the wild in 2000. The Werribee Open Range Zoo near Melbourne, Australia, and other facilities are now breeding the oryx with the hope that individuals can be reintroduced into the wild.

According to the World Association of Zoos and Aquariums, over 700 million people worldwide visit zoos and aquariums each year. In the United States, more people visit these attractions than attend major league football, basketball, hockey, and baseball games combined. Clearly, there is great potential for zoos to educate and contribute to conservation. Personal development of strong conservation ethics just might begin with a childhood visit to the zoo.

JULY 2

Nature's Phenology

Depending on where you live, early July might bring monsoon rains and hot temperatures, or cold and dry conditions. Frogs might be breeding, or aestivating. Flowers might be blooming, or wilting. We pay attention to the phenology of plants and animals—the cycles of their activities relative to climatic conditions. We rejoice at seeing the first buds on our fruit trees, the first violets blooming in the woods, and the first ripe blueberries. We notice when maple leaves turn flaming colors, when squirrels begin to hoard nuts, when insects become scarce, and when the first snowflakes fall.

Scientists are documenting changes in the phenology of plants and animals to understand how they respond to climate change. In areas where temperatures are warming earlier in the spring, frogs are calling earlier, birds are nesting earlier, and leaf and flower buds are opening earlier. In areas where temperatures are cooling off later in the autumn, snakes are going into hibernation later, leaves are falling later, and mosquitoes are breeding later.

Why not help scientists accumulate observations of nature's phenology? If you live in the UK, join "Nature's Calendar" online and submit your observations. If you live in the US, join "Nature's

Notebook." Many other citizen science programs around the world offer similar opportunities to help scientists study plants, animals, and climate change. Your senses of sight, hearing, and smell will be heightened as you become even more aware of nature's phenology.

JULY 3

Crushed by a Boot

Great auks must have been magnificent birds, dressed in black and white plumage, sporting huge, hooked bills, and standing nearly as tall as human toddlers. These flightless birds lived in the cold North Atlantic coastal areas of Canada, the northeastern United States, Norway, Greenland, Iceland, the Faroe Islands, Ireland, Great Britain, France, and northern Spain. Fishermen strangled the last confirmed pair of great auks, at Eldey Island, Iceland, on July 3, 1844, killed for a merchant who wanted specimens to sell to collectors. In the rush to grab the adults, one fisherman stepped on the last auk egg.

Most of what we know about great auks comes from sailors' accounts, as the birds were never the focus of scientific study. Great auks lacked an innate fear of people. That, combined with their inability to fly and clumsy waddling on land, made them vulnerable to exploitation. Auks were killed as food, their flesh used as fishing bait, their beaks coveted as decorative items, their down used for pillow stuffing, and their eggs collected for food and displayed as curiosities. They were stamped out over 170 years ago by human greed and overharvesting.

The American Ornithologists' Union, founded in 1883, named its journal *The Auk*, published since 1884. By honoring this bird and raising its prominence in the public eye, the founders hoped to draw attention to the consequences of human overexploitation.

Hum, Whistle, Crackle, and Bang!

The Fourth of July (Independence Day) in the United States celebrates the adoption of the Declaration of Independence in 1776. Fireworks accompanied the first anniversary celebrations in 1777, and they have added a magical touch to the holiday revelry ever since. Chemicals react and produce displays of many-colored lights and noises, ranging from agreeable hums, whistles, and crackles to deafening bangs. Fireworks date back to the seventh century CE, invented by the Chinese, who set them off during festivals to frighten away evil spirits.

Humans aren't alone in using chemical reactions to produce explosions. More than 500 species of bombardier beetles live on all continents except Antarctica. These ground-dwelling beetles spray hot, noxious chemicals from their rear ends, not to frighten away evil spirits, but to repel predators such as ants. A bombardier beetle has two glands in its abdomen, each consisting of an inner and an

outer compartment. When the beetle senses a threat, it squeezes hydrogen peroxide and hydroquinones from the inner compartments into the outer compartments. There, catalysts (enzymes) trigger an instant explosion. Heat from the reaction brings the chemical mixture to near the boiling point of water, and the pressure of oxygen gas causes the mixture to blast out the beetle's rear end. Nature's own chemical explosions are just as magical as the human-made.

JULY 5

Dolly the Sheep

> Dolly [the sheep] is derived from a mammary gland cell and we couldn't think of a more impressive pair of glands than Dolly Parton's.
> —SIR IAN WILMUT, one of the scientists who cloned Dolly the Sheep

Dolly, the first mammal cloned from an adult cell, was born in Scotland on July 5, 1996. She began with a cell taken from the udder of a Finn Dorset white sheep, but Dolly had two additional mothers. A Scottish Blackface ewe provided the de-nucleated, unfertilized egg into which the udder cell nucleus was injected. After six days in a test tube, the embryo was implanted into a surrogate mother, another Scottish Blackface ewe. Dolly died of lung disease at the age of six and one-half years. Her remains are on display in the National Museum of Scotland in Edinburgh. She captured the imagination of the world, allowing science fiction to become reality—a cell grown into a carbon copy of its original donor. After Dolly, scientists successfully cloned various other large mammals, including horses, bulls, deer, and pigs.

Some scientists advocate cloning endangered species to avert extinction. Dolly showed us that, theoretically, it might be possible to do this. Other scientists argue that cloning doesn't address the problems that caused declines in the first place. To save endangered species, they say we should use our limited resources to mitigate the root causes of declines, such as overhunting, habitat modification and destruction, and introduced species. Still other scientists suggest we should implement both approaches.

JULY 6

Filial Piety

The white stork symbolized filial piety for the ancient Greco-Roman world, where mythology portrayed the birds as caring for their aged parents. The association was not based on natural history, but perhaps reflected the birds' gregarious behavior. Aristotle (384–322 BCE), Pliny the Elder (23–79 CE), and Aelian (ca. 175–235 CE) praised the stork's devotion. Artists depicted a flying stork carrying its parent on its back, feeding the parent with a large fish. Later, Edward Topsell (ca. 1572–1625), English author of a well-known bestiary, declared the stork "a bird of God" for its virtues.

Europeans have long considered it good luck to have storks nest on their rooftops. The Swiss naturalist Conrad Gesner (1516–1565) and others claimed that after a stork built its nest on a rooftop, it left

behind one of its young as thanks to the homeowner. This belief led to the widespread European folklore that the stork delivers human babies in baskets, a tale popularized by Hans Christian Andersen's story *The Storks* (1838). In more recent times, birth announcements feature storks carrying human babies in pink or blue blankets.

White storks are still much appreciated and honored with spring and summer festivals. Stork celebrations often include constructing nesting platforms and street dancing to celebrate these pious and generous birds, heralds of new life. During the first weekend in July, Slovenia hosts Stork Days, a special place to celebrate because the white stork is a symbol for the region.

JULY 7

Food of the Gods

Chocolate, made from the seeds of the cacao tree *Theobroma cacao*, can be intoxicating. The generic name comes from the Greek words *theos* and *broma* meaning "food of the gods." July 7 is World Chocolate Day. Indulge! Medical experts tell us that chocolate may lower cholesterol levels, reduce risk of cardiovascular problems and stroke, and slow memory decline. Some historians suggest it was on this day in 1550 that chocolate as a drink was introduced in Spain.

Christopher Columbus returned to Spain in 1504 with cacao beans, but they were ignored in favor of other New World treasures. Spanish explorer Hernán Cortés learned about chocolate when he visited the Aztec emperor of Mexico, Montezuma II, in 1519. The Aztecs drank *xocoatl*, made from cacao beans ground into a paste, flavored with vanilla and chili pepper. Montezuma supposedly drank several goblets of *xocoatl* to increase his libido before visiting his harem. Cortés took cacao beans with him when he returned to Spain in 1528, but some historians suggest that it wasn't until 1550 that the Spanish began to drink chocolate. The Spanish kept chocolate a secret until the daughter of King Philip of Spain married Louis

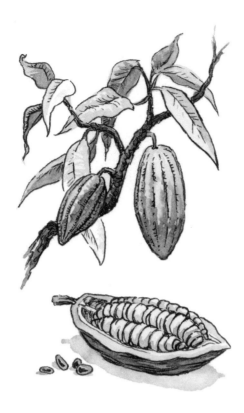

XIII of France in 1615 and took her love of chocolate with her. From France, chocolate spread across Europe.

Eating chocolate increases the brain's level of serotonin, and the phenylethylamine (PEA) in chocolate triggers release of endorphins. These natural mood-altering chemicals make you feel happy and are just one more reason to enjoy some "food of the gods" today.

JULY 8

Music with the Fish

The second Saturday in July sees the annual Lower Keys Underwater Music Festival at Looe Key Reef, Florida, south of Big Pine Key. Looe Key is touted as one of the most beautiful reefs in the Florida Keys. The festival was founded in the mid-1980s to raise awareness for coral reef protection. Coral reefs are the "tropical rainforests of the sea," an epithet that refers to their high biodiversity. Although coral reefs occupy less than 0.25 percent of the ocean floor worldwide, they provide a home for one-quarter of all species of marine life.

Join several hundred other divers and snorkelers to explore the only living coral barrier reef in the Continental United States, while swaying to the Beatles' "Yellow Submarine" and "Octopus's Garden," Jimmy Buffett's "Fins," humpback whale songs, and other ocean-themed music streamed from underwater speakers. Experience sound traveling over four times faster than in air, for a sonic experience that has been described as "particularly ethereal." Swim with yellow, orange, red, blue, and purple fish that inhabit the reef. Watch fanciful mermaids and mermen lip-synching and pretending to play whimsical instruments, including French angelfish horn, fiddle crab, harmoni-crab, trombone fish, and drumfish, sculpted by local artists. Come in costume and compete for the "best-dressed" prize. Most of all, come and appreciate the beauty and diversity of the spectacular living architecture and inhabitants of Looe Key Reef.

JULY 9

Sweet Grass

During the mid-1700s, Moravian settlers in Nazareth, Pennsylvania, created "Nazareth cookies," forerunner to our modern-day sugar cookies. In 2001 Pennsylvania declared the Nazareth sugar cookie its official cookie, but for all of us July 9 is National Sugar Cookie Day.

The granulated sugar in your cookies likely comes from sugarcane, a grass native to tropical regions of South Asia and Melanesia. Sugarcane is thought to have been cultivated in New Guinea some 8,000 years ago. From there, cultivation spread to other South Pacific islands; within 2000 years sugarcane was cultivated in Indonesia, the Philippines, and northern India. Today, sugarcane is grown worldwide in subtropical and tropical climates. About 80 percent of the world's sugar comes from sugarcane, with most of the rest from sugar beets.

Sugarcane has shaped human history. Africans were taken as slaves to work in sugarcane plantations in the Caribbean and elsewhere; Indians, Chinese, and Portuguese migrated across the globe to work as indentured laborers in cane fields. Sugarcane has also shaped ecosystems through conversion of the landscape, and by modifying plant and animal communities. Cane toads were introduced into Florida, Hawaii, Puerto Rico, the Philippines, and Australia to control cane beetles that eat sugarcane roots. Mongoose were introduced into Hawaii and Jamaica to control rats in sugarcane fields. Cane toads and mongoose are examples of biological control gone awry, as the introduced species continue to wreak havoc in their adopted ecosystems through competition, predation, and other negative interactions with native fauna. In 2008, the Better Sugar Cane Initiative was established, which sets global standards to certify sustainable production and reduce the environmental impact of sugarcane production.

JULY 10

Wilderness Protection

> Wilderness: An area where the earth and its community of life are untrammeled by man, where man himself is a visitor who does not remain.
> —Legal definition of wilderness, The Wilderness Act of 1964

The Boundary Waters Canoe Area in northeastern Minnesota has seen vast changes since ancient glaciers exposed bedrock and formed a network of lakes and streams. During the first half of the nineteenth century, fur traders so exploited the area's furbearing mammals that trappers were forced to move farther west. Before the end of the nineteenth century, iron ore had been discovered. By 1895, a campaign was begun to preserve the area for future generations. In 1909 President Theodore Roosevelt officially designated over one million acres as Superior National Forest.

Conflicts arose, however, between those who believed the Superior National Forest should be used for recreation, and those who advocated timber harvesting and generation of hydropower. On July 10, 1930, at the urging of conservationists, President Herbert Hoover signed into law the Shipstead-Newton-Nolan Act, the first US statute declaring that land be protected as "wilderness." The law was designed to protect water levels and lakeshores for recreational use by prohibiting dams and logging within 400 feet (122 m) of recreational waterways in the Canoe Area. Wilderness was viewed as essential for the wellbeing of the human spirit.

In 1964, President Lyndon B. Johnson signed the Wilderness Act into law. The act created the legal definition of wilderness and protected 9.1 million acres of federal land as wilderness. Today more than 100 million acres are protected within the US Wilderness System.

JULY 11

The "Magic" of Metamorphosis

As children, we are fascinated by frog metamorphosis. Starting off as pouty-mouthed, swimming blobs powered by finny tails, tadpoles transform into big-mouthed predators hopping about on strong legs. How did the pouty mouth become so wide? What happened to the finny tail? And where did the legs come from? It is all a mystery that seems magical. Later, in high school biology class, scientific explanation supplants magic.

On July 11, 1912, J. F. Gudernatsch, an anatomist from Cornell Medical School, published a paper reporting that metamorphosis

in tadpoles is controlled by thyroid hormones. He found that when he fed tadpoles mashed-up thyroid glands, they quickly developed legs, absorbed their tails, and precociously transformed into tiny frogs. We've learned a lot about amphibian metamorphosis since Gudernatsch's experiments, but his work formed the basis of our understanding. Tadpole development involves two processes: growth (increase in size) and differentiation (structural changes in the tissues). Different hormones control these two processes. Hormones secreted by the pituitary gland largely regulate growth, whereas thyroid hormones and corticosteroids (produced by the adrenal glands) regulate differentiation.

Hormones aside, metamorphosis still seems a bit magical. Worldwide, frogs still symbolize rebirth, renewal, and resurrection because of their transformation, just as they did for the ancient Egyptians, Romans, and Chinese; the Olmec of Mexico; and the Egyptian Copts, leaders in developing the early Christian church.

JULY 12

Thank a Cow

The second Friday of July is Cow Appreciation Day in the United States, a day set aside to remind us how much we depend on cows.

They give us much more than milk and beef. Gelatin made from boiling skin, tendons, and bones is used in making marshmallows, gummy candy, some ice creams, shampoo, cosmetics, and vitamin capsules. Hides become jackets, belts, shoes, purses, and footballs. Fat is used in making soap, leather conditioners, and dynamite. Bone is cut into piano keys and buttons. Intestinal fibers become tennis rackets and cello strings. Cow manure is a great organic fertilizer, and dried cow dung can be used as fuel. Hooves become the treats you give to Fido, and fine hairs from ears and tails are made into "camel hair" paintbrushes.

While most of us appreciate cows for their products and by-products, Hindus consider cows sacred. These domesticated herbivores provide life-sustaining milk. Cows are always giving, and they take nothing but water, grass, and grain; they are gentle, yet strong. Hindus honor cows at festivals, decorating them with garlands and offering special feedings. Cows symbolize life itself.

Depending on your personal persuasions, you could celebrate the day with grilled steaks or barbequed ribs. Or, you could abstain from using soap, washing your hair, eating marshmallows, or treating Fido with rawhide chews.

JULY 13

Ra

Ancient Egyptians claimed July 13 as the day their sun god Ra was born. Ra was king of the gods, patron of the pharaoh, and creator of everything. He created himself from Nun, the preexisting void. Then, from spittle he created air and moisture, the parents of Sky and Earth. Ra castrated himself, and humans arose from the drops of his blood. Ra's symbols were the falcon, serpent, and scarab (dung) beetle. The god was often depicted as a man with the head of a falcon, above which sat the sun disk encircled by a cobra. He was also depicted as a dung beetle rolling the sun across the sky, just as

the scarab rolls its ball of dung.

Turtles and a huge serpent named Apep, all of whom lived in the Underworld, were Ra's enemies. Every night as Ra journeyed through the aquatic Underworld on his nightly 12-hour voyage, the turtles and Apep tried to prevent the god's passage in a perpetual war of darkness against the forces of light. But every morning, the rising sun revealed that Ra had conquered the turtles and Apep and renewed the world. Ancient Egyptians so hated turtles and so loved Ra that they would chant, "May Ra live and may the turtle die."

Although most of us today don't literally worship the sun, we are still utterly dependent on its light and heat energy. Happy Birthday, Ra!

JULY 14

Show No Disrespect

Today is Shark Awareness Day in the US. An estimated 100 million sharks are killed each year for their flesh and fins; countless others are caught unintentionally by fishers targeting other species. Shark Awareness Day highlights the value of these keystone marine predators, and the need to protect them. Sharks are far more threatened by us than we are by them.

People whose lifestyles allow them the luxury of wading into the ocean, or not, often fear sharks as menacing creatures—better dead than alive. In contrast, people of some cultures who live closely with sharks accept them as forces over which they have no control, much like typhoons and volcanic eruptions. Some see sharks as protectors, spiritual beings, and even gods. Polynesians on Molokai in the Hawaiian archipelago revered the shark god Kauhuhu, and Fiji Islanders had a shark god Dakuwaqa. Solomon Islanders erected stone altars and offered human sacrifices to their shark god Takw Manacca. Peoples of Papua New Guinea, Hawaii, and French Polynesia viewed sharks as reincarnated ancestors.

Marshall Islanders have long recognized the constructive role sharks play in the health of their islands and the sea. In the past, Marshall Islanders fought each other when one tribe showed disrespect to the sacred shark of another tribe. People believed that if fishermen respected sharks, the fish would protect them; stories are told of sharks guiding lost fishermen to shore. Today the Marshall

Islands are home to the world's largest shark sanctuary—a safe haven where shark fishing is banned in an area of the Pacific Ocean two-thirds the size of the continental United States.

One Thousand Cranes

Obon, the Buddhist Festival of Souls, begins in Japan on either July 15 or August 15, depending on the region. The festival lasts for three days, during which the spirits of one's ancestors are said to revisit the household altars. It is a time to honor the ancestors' spirits by visiting and cleaning gravesites and to commemorate the deceased with song and dance. Some offer 1000 origami cranes.

Cranes symbolize longevity, happiness, and good fortune for the Japanese. According to ancient Japanese legend, anyone who folds 1000 origami cranes will be granted a wish by the gods; the wish can be passed on to someone else. The number 1000 derives from the folk belief that cranes live 1000 years. One thousand origami cranes, often assembled as 25 strings of 40 cranes each, are given as a traditional wedding gift or presented to a newborn baby, representing wishes for a long, happy life.

Cranes also have become a symbol of peace. Several temples in Hiroshima have eternal flames representing everlasting peace. Visitors leave strings of 1000 origami cranes at the temples to support the prayer for world peace. Every year during Obon, people leave strings of 1000 origami cranes at the Hiroshima Peace Park in memory of their ancestors' spirits and as a wish for world peace. The cranes are often left exposed to the sun, rain, and wind. They become tattered bits of red, pink, yellow, orange, green, blue, purple, white, and tan, blowing in the breeze like prayer flags, as the wishes are released.

Good and Evil

July 16 is World Snake Day, a time to educate the public about these often-persecuted reptiles. Snakes have long been seen as both good and evil, deity and devil, healer and killer, life and death, creation and destruction.

People from diverse cultures fear and hate snakes. In Indian mythology, the snake demon Vritra swallowed all the world's water and caused a devastating drought. Serpents represented chaos in Egyptian mythology. Greek and Roman myths depicted snakes as villains to be dispatched. The serpent symbolized evil in the Garden of Eden. In a 2001 Gallup poll of adult Americans, 51 percent of respondents said their greatest fear was snakes. We fear what we don't understand and that which is different from us. Snakes have no arms or legs; they slither on their bellies. They swallow their food whole. They stare. (They can't help it—they have no eyelids and cannot blink.) They have forked tongues, and they literally crawl out of their old skins.

At the same time, cultures worldwide and throughout time have associated snakes with wisdom, health and healing, resurrection and rebirth. In both the New and Old Worlds, people wear snake amulets, charms, and talismans for their perceived powers to provide and protect. Cobras are revered in India and Pakistan, and pythons are worshipped in Benin, western Africa. Worldwide, people value snakes for their capacity to control rodents. If you find snakes fascinating, today is a good day to share your view with those who fear them.

The Inca and the Llama

Four thousand years ago, the Inca who lived in the Andes near Lake

Titicaca domesticated llamas. They sacrificed llamas to their gods, wove llama fiber into fabric, ate their meat, and used parts of their digestive tract as medicines. By 680 BCE, the Inca had begun to use llama manure to enrich the soil. The rich fertilizer allowed the Inca to grow corn at these high elevations. Historians speculate that the ability to grow corn enabled the Incan civilization to spread and ultimately conquer much of South America.

During the Inca Empire (ca. 1425–1535 CE), the world's largest state at the time, the Inca used llamas as beasts of burden on their extensive mountain road system. Llamas were their most valued domesticated animals. Gold was the emblem of their sun god, representing the sun's powers of regeneration. No wonder, then, that they made llama figurines from gold, which they offered to their gods. When the Spanish invaded in the late 1520s, they recorded the abundance of these gold llama figurines. Few remain today. Most were melted down for bullion and sent to Spain.

Now in the twenty-first century, llamas are valued in many parts of the world, still raised for wool and meat. The ancient Inca might smile to see llamas carrying today's hikers' camping gear during "llama treks." Llamas are celebrated with food, music, and dancing during Llama Fest, hosted in Spanish Fork, Utah, on the third Saturday in July.

Passionate about Birds

> When nature made the bluebird, she wished to propitiate both the sky and the earth, so she gave him the color of one on his back and the hue of the other on his breast, and ordained that his appearance in spring should denote that the strife and war between these two elements was at an end.
>
> —JOHN BURROUGHS, "The Bluebird"

"What do you want to be when you grow up?" is a question we commonly ask children of any age. Early in life, many of us flit from passion to passion, extending our intellectual tentacles to answer this question. Encouragement and opportunity often guide and ultimately cement our life's path.

For teenagers passionate about birds, the Cornell Lab of Ornithology, located in Sapsucker Woods just outside Ithaca, New York, offers a four-day workshop between mid-July and early August—the Young Birders Event. The workshop provides the opportunity to meet and interact with professionals working with birds, including ecotour leaders, environmental educators, policymakers, and academic ornithologists. Activities include a sound recording workshop, an eBird and field notes workshop, two days of birding field trips, and presentations.

John Burroughs (1837–1921), preeminent American naturalist and a master at nature-writing, expressed a special fondness for birds. He would have been an inspirational professional, offering a complementary perspective, for the teenagers attending the Young Birders Event today.

JULY 19

Happy as a Clam

They go back 510 million years. They live in both freshwater and marine habitats. They range in size from nearly microscopic to more than 47 inches (120 cm) across. They spend most of their lives buried in the sand. We're talking clams—mollusks with two shells joined together by a hinge joint and ligament.

Some clams are good to eat—raw, steamed, boiled, baked, fried, or roasted. Popular dishes include baked stuffed clams, linguine with clam sauce, steamed clams in butter and sake, clam creole, clam curry, clam cakes, clam pizza, risotto with clams, Manhattan and New England clam chowder, clam fritters, and clam dip.

The Yarmouth Clam Festival in Yarmouth, Maine, has been held since 1965. Today the festival is a fundraiser for about 40 local nonprofit organizations, school groups, and churches. Come and join the more than 100,000 annual visitors for the festival the third weekend in July. The three-day event hosts live music, a juried craft show, carnival, clam-shucking contest, canoe and kayak race, fireworks display, and a parade with floats, marching bands, and antique cars. During the festival, over 6000 pounds of clams are served, in addition to 2500 pancake breakfasts, 6000 lobster rolls, 6000 strawberry shortcakes, and 13,000 lime rickeys. It's a long weekend to feel "happy as a clam at high tide."

JULY 20

Bunny Girl

On July 20, 1969, Apollo 11 landed on the moon, and Neil Armstrong and Buzz Aldrin walked on its surface. I was doing fieldwork in Belém, Brazil, at the time. I listened to the broadcast at the American consulate, along with about 25 other Americans, the governor

of Pará, and other Brazilian dignitaries. As Apollo 11 landed, we all burst into shouts and shared bear hugs. We raised the American flag outside and enjoyed several rounds of champagne.

The following day, my Brazilian assistants asked me, "How much of the moon will the Americans take?" They had heard that the astronauts had collected samples of rocks and soil, and they feared that Americans would claim the moon for themselves. I explained that two years earlier more than 90 nations, including the United States, had signed a space exploration treaty stating that no country could own, or use for military purposes, any natural body in outer space, including the moon. The moon belongs to everyone. It belongs to no one.

Many human cultures see images in the moon. People from Western cultures often imagine a man in the moon. Folklore from Africa, China, and First Peoples from North America tells of a frog or toad in the moon. Pre-Columbian peoples from Mesoamerica associated a rabbit with the moon, as do Chinese, Japanese, and Koreans. Reportedly, Aldrin looked for, but did not see, "the bunny girl" (his words) on the moon.

JULY 21

Florida Reefs

The third week in July is Reef Awareness Week, a celebration of the coral reefs that parallel the Florida Keys. Florida's coral reefs are 10,000 years old, formed when sea levels rose after the last Ice Age. The reef architecture is composed of about 100 species of corals, including brain coral, gorgonians, star coral, elkhorn, staghorn, sea fans, and sea whips. The reef provides apartment lodging for sponges, anemones, sea cucumbers, starfish, clams, and more. At least 1000 species of fish live in or around the reefs, eating from a veritable smorgasbord: clownfish eat zooplankton; angelfish munch on sponges; butterflyfish eat coral polyps; queen triggerfish eat sea urchins; green moray eels feast on crustaceans; and sharks, barracudas, and groupers prey on other fish. Parrotfish graze on algae growing on the corals; they also grind up some of the corals' calcium carbonate skeletons—and poop it out as white sand.

Corals reefs worldwide are threatened by pollution, warming ocean temperatures, coral disease, and coral bleaching (loss of endosymbiotic algae). Scientists have recently found that Florida's coral reefs are disintegrating faster than originally predicted, driven by accelerating climate change. As the climate warms, more atmospheric carbon dioxide is absorbed by the ocean water, forming carbonic acid, which interferes with the corals' ability to make their skeletons. If the coral reef ecosystem dies, other marine life will die as well. Reef Awareness Week focuses on environmental education. One of the most important ways we can help protect these ecosystems is to reduce fossil fuel emissions, which lead to ocean warming. We can do this by walking, biking, or riding the bus instead of driving our cars.

JULY 22

Alligators, Waterfalls, and Magnolias

His enormous body swells. His plaited tail, brandished high,
floats upon the lake. The waters like a cataract descend
from his opening jaws. Clouds of smoke issue from his dilated

nostrils. The earth trembles with his thunder.
—WILLIAM BARTRAM, *Bartram's Travels*

William Bartram was America's first well-known, native-born naturalist/artist. In March 1773, Bartram sailed from Philadelphia and began a 4-year journey through the southern colonies (current states of North and South Carolina, Georgia, Florida, Alabama, Mississippi, Louisiana, and Tennessee). He dried plant specimens, sewed them into linen books, and transported them on pack animals. He took careful notes and made detailed sketches of plants and animals, Native Americans, and landscapes. In 1791, Bartram published a description of his travels and the natural history he encountered in a 4-volume book whose 50-word title is usually shortened to *Bartram's Travels*.

Bartram represented a fresh style of natural history writing for North America: portrayal of nature through personal experience interwoven with scientific observation (rather than purely scientific description). His portrayal, above, of an alligator—the first he had ever seen—is one such example. Bartram's writings of the exotic, subtropical region were so alluring that the Romantic poets William Wordsworth and Samuel Taylor Coleridge borrowed his vivid imagery for their poetry: Wordsworth in "Ruth," and Coleridge in "Kubla Khan."

Bartram died on July 22, 1823, while strolling in his beloved garden in Philadelphia. The garden still blooms today, more than 250 years after his father, John Bartram, planted it.

JULY 23

Water is Life

Thousands have lived without love, not one without water.
—W. H. AUDEN, English poet

Newborn human babies are about 75 percent water. By the time we reach adulthood, we're 55–60 percent water. Trees are generally 60–80 percent water. All living organisms need water to survive, but water is more than a critical resource. It has a spiritual component that connects us to nature. We listen to simulated sounds of the ocean's waves to relax and fall asleep. We skip stones, swim, fish, and canoe in water to release stress.

Water is associated with cleansing and healing, and thus is often a focus in renewing celebrations. In Myanmar, Cambodia, Thailand, and Laos, people pour water on each other as part of a New Year's tradition that reflects symbolic cleansing to prepare for the new year. In many South American countries during Carnival, celebrated two weeks before Lent, people throw water balloons at each other to purify the body.

Armenians celebrate a water festival called Vardavar. Originally, Vardavar was celebrated during mid-July in honor of Astghik, pagan goddess of water, love, and fertility. Ancient legend tells that Astghik spread love over Armenia by pouring rose water. Armenians honored the goddess by sprinkling water infused with rose petals on each other. Later, the pagan celebration evolved into a Christian holiday called the Feast of the Transfiguration, often still called Vardavar. Today, Armenians dump water on each other during the July Vardavar festival to convey wishes of good health, to purify and bestow a fresh start. Water is life.

JULY 24

Gathering in the Highlands

Golden retrievers generally rank between the third and seventh most popular dog breed in the United States in any given year (Labrador retrievers have ranked #1 for over a quarter-century). Goldens are loved for their loyalty, sweet and friendly disposition, and intelligence.

Late July sees the three-day Golden Retriever Festival in the highlands of Scotland, at the Guisachan Estate. It was on the grounds of this estate that businessman and politician Lord Tweedmouth founded the breed in 1868. Tweedmouth loved waterfowl hunting, and the existing breeds weren't good at retrieving from both land and water. He began by crossing a yellow wavy-coated retriever with a tweed water spaniel (a now-extinct breed known to be an eager retriever). The four resulting puppies began a breeding program that included Irish setter, bloodhound, St. John's water dog, and black retrievers.

The first gathering to pay tribute to the man who began breeding golden retrievers took place in 2006. People come to Scotland from as far away as Japan, Australia, and North America. The gathering in 2018 celebrated the 150th anniversary of the founding of the breed. During the festival, participants play tug-of war games, pitting nation against nation, and hurl frozen haggis (after donning a kilt and downing a shot of whiskey). And, of course, there's a one-breed dog show.

JULY 25

Purple Spikes

Bright purple flowers poke through the forest undergrowth above Silverton, Colorado—a patch of larkspur, 10-inch (25-cm) spikes of regal purple flowers, each with four petal-like sepals and a fifth sepal resembling the long claw on a lark's toe. Larkspur is also called delphinium, a common name for all species in the genus *Delphinium*, both wild and cultivated.

All parts of the delphinium plant contain alkaloids, which render them poisonous to humans and livestock. According to Italian lore, after warriors slew a dragon, they wiped their swords on the ground. Poisonous blue delphinium flowers grew from the dragon's blood. During the first century CE, the Greek physician Dioscorides

recommended drinking delphinium seeds in wine to cure scorpion bite, based on the belief that delphinium leaves can be used to paralyze scorpions. Also during the first century CE, the Roman naturalist Pliny the Elder recommended a tincture made from delphinium seeds for topical use to heal wounds and to kill adult lice and nits in the hair. In Transylvania, delphinium was planted to keep witches away from horse stables. In England, it was believed that witches commonly included delphinium in their toxic brews.

Larkspur (*Delphinium*) is the birth flower for July. Never mind that the plant contains alkaloids to protect itself. Do we denigrate the rose for its thorns? Larkspur symbolizes love and joy, with certain colors associated with specific qualities: pink for fickleness, white for happiness, blue for loyalty, and purple for first love.

JULY 26

Animal Folktales

The fourth week in July sees the Folktales Festival Greece, in Kea, Greece. You'll find an abundance of animal-oriented folklore—not because the festival specifically focuses on animals, but because of the nature of folklore itself. We have developed alternative perspectives of animals through folklore, reflecting the close link between people and other animals. Folklore portrays animals as creators, destroyers, reincarnated ancestors, advisors, teachers, guardians, messengers of the gods, and as gods themselves. Animals delight us as tricksters, inspire us as culture heroes, and guide us as role models.

Storytelling has a long history in Greece. One of the most widely known storytellers from Greek antiquity was Aesop (ca. 620–564 BCE), a slave who won his freedom by telling fables—animal tales that convey morals and illustrate truths. (Some experts suggest that there was no such person, and that what we call "Aesop's fables" were told by several people.) From Aesop's "The Ant and the Chrysalis" we learn that appearances are deceptive, and from "The Fox and the Grapes" that it is easy to despise that which is beyond our reach. "The Bald Man and the Fly" teaches that revenge hurts the avenger, and "The Hares and the Frogs" shows us that there is always someone else worse off than ourselves. Aesop's animal folktales teach societal values and appropriate behavior without being preachy. They illustrate universal truths that people everywhere recognize and value.

JULY 27

Ukai

A fisherman tethers the base of his bird's throat to keep it from swallowing the big ayu ("sweetfish"). He signals, and watches the bird plunge into the water. Once the bird catches a big one, he gently pulls the bird by its reins back to the boat and eases the fish from its crop (the pouch that holds food before it heads to the stomach). *Ukai*, the art of fishing for ayu with trained cormorants, has been practiced for 1300 years in Japan. Typical maximum size of ayu is about 12 inches (30 cm). Fishing is done at night, by the light of a

small pinewood fire in a basket hanging from a pole projected over the front of the boat. The fire serves both to light the way and to attract ayu.

It takes about three years to train Japanese cormorants for *ukai*. Fishing masters profess a close partnership with their birds, stroking their heads and bellies daily to maintain trust. The birds receive occasional fish rewards, but they also swallow the smaller ayu. Fortunately for the fishermen, the cormorants haven't figured out to catch only the small ones.

If you visit Gifu City, Gifu Prefecture (Honshu island), Japan, any night from mid-May through mid-October (except during harvest moon, when the river level is too high), you will see *ukai* on the Nagara River. English comic actor Charlie Chaplin watched *ukai* twice on this river and called it "the finest art Japan has to offer." Or, visit Yamanashi Prefecture (Honshu island) any Thursday through Sunday from July 20 to August 19 for the Ukai Cormorant Fishing Festival on the Fuefuki River.

JULY 28

Paleontological Supremacy

Many of us are driven by the challenge to be the best we possibly can, whether a superb violinist, track star, pediatrician, high school math teacher, mother, father, or some other worthy endeavor. Scientists thrive on the challenge to discover new species, chemical elements, or planets; propose enduring theories; or explain processes or phenomena.

Edward Drinker Cope, born on this day in 1840, was a driven scientist. His inner challenge was to best his rival, fellow paleontologist Othniel Charles Marsh. A period of time now known as the "Bone Wars," between 1877 and 1892, saw Cope (from the Academy of Natural Sciences in Philadelphia) and Marsh (from the Peabody Museum of Natural History at Yale University) compete to discover the most species of dinosaurs. Each dug and blasted his way across the western US, engaging in bribery, theft, and destruction of bones and fossil sites to prove himself the preeminent paleontologist and to portray his rival as incompetent. Between the two of them, they discovered and described more than 140 new species of dinosaurs, including *Allosaurus*, *Apatosaurus*, *Stegosaurus*, and *Triceratops*, though only 32 of these species remain valid today. (Many of their named species were not truly new or unique.) Cope published more than 1400 scientific papers, a record still held today, and he named over 1000 vertebrate species. He lives on in the scientific journal named after him, *Copeia*, the journal of the American Society of Ichthyologists and Herpetologists.

JULY 29

The Pungent Bulb

Garlic is very potent and beneficial against ailments caused by changes of water and of location. It drives away snakes and scorpions by its smell and, according to some authorities, every kind of wild beast.

—PLINY THE ELDER, *Naturalis Historia* (77–79 CE)

Pliny the Elder may have exaggerated the protective benefits of garlic against dangerous animals, but the beliefs persisted for hundreds of years. Central Europeans hung garlic in their windows and rubbed garlic on their chimneys and keyholes to protect against vampires, werewolves, witches, and evil spirits.

Ancient Egyptians, Chinese, Indians, Greeks, Romans, and others used garlic for healing a range of ailments, and we still use the herb medicinally. Garlic contains allicin, a natural antibiotic. During both World Wars, garlic was used as an antiseptic to prevent gangrene and other bacteria-caused wound infections. Recent studies have suggested that garlic preparations may lower total cholesterol and that garlic consumption is associated with a lower risk of stomach cancer. No doubt we will continue to discover medicinal benefits of the pungent bulb.

The last full weekend in July sees the Gilroy Garlic Festival in Gilroy, California, founded in 1979 and run as a fundraiser for local charities. Festival-goers can participate in the garlic-braiding

workshop and Great Garlic Cook-off; watch garlic-cooking demonstrations; and sample garlic-laced calamari, salmon, hot wings, and French fries. For the adventurous, there's garlic-flavored ice cream and chocolate.

JULY 30

A Prickly Issue

The United Kingdom has the lion as its national animal. In 2013, BBC Wildlife invited the public to vote on 10 species of plants and animals for consideration as another UK national species: badger, bluebell, hedgehog, ladybug, oak tree, otter, red squirrel, robin, swallow, and water vole. On July 30, 2013, the winner of 9108 votes cast was announced: the hedgehog, with 3849 votes.

Many Brits love hedgehogs because of Beatrix Potter's Mrs. Tiggy-Winkle, and it doesn't hurt that they devour slugs, caterpillars, and other garden pests. Unfortunately, however, people threat-

en hedgehogs through habitat modification. Whereas in the 1950s there were an estimated 36.5 million hedgehogs in Britain, by 2015 the number had declined to fewer than 1 million. Hedgehogs are long-distance commuters, sometimes roaming over a mile (1.6 km) each night. Their wandering takes them across roads, leading to an estimated 50,000 hedgehogs run over by vehicles each year. Advocates for the hedgehog note that if the hedgehog becomes a national symbol, it might earn the right to have its habitat protected.

Not everyone thinks the hedgehog would make an honorable national animal. In November 2015, Rory Steward, the Environment Minister, spoke on the issue to the House of Commons. He asked whether the UK wants as its national animal one that sleeps for six months out of the year, and, when confronted with danger, rolls into a hissing ball of spines. As of August 2017, the hedgehog has not yet been ratified to take a place alongside the lion.

JULY 31

Survivors and Anomalies

On July 31, 2015, BBC News reported on a more than 2200-year-old *Macrotermes falciger* termite mound (abandoned in recent decades) from the Democratic Republic of Congo. The news spread quickly because of the astounding discovery that termites can live in the same mounds for millennia. Species in the genus *Macrotermes* build the most complex structures of any insect—large mounds consisting of a maze of temperature and humidity-controlled passages. Mounds constructed by *Macrotermes falciger* can reach 30 feet (9 m) in height and more than 50 feet (15 m) in diameter.

Although termites get a bum rap, most species don't deserve it. On a geological timescale, termites have been around a long time. They lived alongside the dinosaurs, and they are still with us—over 3000 species strong, found on every continent except Antarctica. On an individual level, queen termites have the longest life span of any insect, some living 50 years. Some are strong enough to eat your home, yet their bodies are soft and delicate. Soldiers defend their colonies, sometimes consisting of several million individuals, yet they can't feed themselves because their greatly modified jaws can't bite into and manipulate food. Termites will never run out of food. They eat mostly cellulose, usually in the form of decaying plant material, in animal feces, or mixed in soil. Only several hundred species do serious damage to our homes and other wooden structures. The rest benefit Earth, as recyclers of dead plant material.

AUGUST 1

Wails of Whales

"Exuberant, uninterrupted rivers of sound," was how biologist Roger Payne described the songs of male humpback whales. In August 1970, Payne released his album "Songs of the Humpback Whale." That same year, folksinger Judy Collins released her album "Whales & Nightingales," on which she sang the Scottish whaling song "Farewell to Tarwathie," accompanied by Payne's whale recordings. The whales' wailings seem to be cries against their inevitable fate. Her

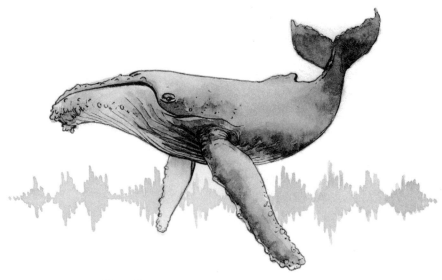

record went gold and introduced millions of people to the songs of these magnificent giants. Payne's album, still the best-selling "sounds of nature" record of all time, helped spawn the Save the Whales movement and led to the 1986 ban on commercial whaling.

Humpback songs, with their recurring phrases, are the most complex songs known of any nonhuman animal. The songs vary each breeding season, with new phrases added and others dropped. During public lectures, Payne often played his cello alongside whale phrases to illustrate the range of the humpbacks' songs.

In 1977, Payne's whale recordings were included on a record carried aboard the Voyager spacecraft, charged with the mission to study the outer solar system. This time capsule also included bird songs, the sounds of surf, wind, and thunder, and spoken greetings in 55 human languages. The sounds were selected to portray the diversity of life on Earth for any intelligent extraterrestrial life forms (or future humans) who might find the record.

AUGUST 2

Down the Rabbit Hole

Alice followed a waistcoat-attired White Rabbit down his hole and tumbled into Wonderland over 150 years ago. It all began in 1862, when Charles Lutwidge Dodgson told a story about Alice and the White Rabbit to amuse a friend's three daughters, as he rowed them five miles up the Thames River. Ten-year-old Alice Liddell asked Dodgson to write down the story. He did. Two thousand copies were printed in June 1865, but Dodgson withdrew the print run because the illustrator was not satisfied with the printing. On August 2, 1865, Dodgson wrote in his diary that "the first 2000 shall be sold as waste paper." *Alice's Adventures in Wonderland*, under the pseudonym Lewis Carroll, was published a few months later.

Many of the animals in the story are anthropomorphized versions of people the Liddell sisters knew, which, for them, must have

made the story especially endearing. Carroll introduced the rest of us to some of the most entertaining anthropomorphic animals in literature, including a White Rabbit who is always in a hurry; a blue Caterpillar who smokes a hookah while sitting on a mushroom and asks Alice if she wants to become smaller or larger; a Cheshire Cat who disappears, but leaves his grin behind; live flamingos used as croquet mallets and hedgehogs as croquet balls; and a sad Mock Turtle, who sings his "Beautiful Soup" song.

Alice's Adventures in Wonderland has never gone out of print, has been translated into 100 languages (including classical Latin), and remains a favorite children's story, a testament to the power of talking animals, fantasy, and the world of brilliant nonsense.

AUGUST 3

Jubilee

August 3, 1952, saw the close of the 15th Olympic Games in Helsinki, Finland. It was the first year that women had been permitted to compete in Individual Dressage (competitive riding in which the horse performs movements in response to signals from its rider). Lis Hartel, representing Denmark, had won the silver medal for Individual Dressage on her horse Jubilee, the first silver medal won by any woman at the Olympics when in direct competition with men.

After the competition the gold medal winner, Henri Saint Cyr of Sweden, carried Hartel from her horse to the Olympic podium. It was one of the most emotional moments in Olympic history, for Hartel was paralyzed from the knees down from polio she had contracted at 23 years old, eight years earlier. Hartel went from being almost completely paralyzed, to learning how to lift her arms, crawl, and eventually walk with arm crutches. Her horse Jubilee helped her resume competitive dressage riding, in spite of Hartel having little muscle control.

Jubilee was calm and dependable and had a quiet temperament, but she had to learn a new routine, including standing motionless while Hartel was lifted on and off her. Jubilee somehow realized that she had to react differently to Hartel, who because she was unable to use her legs, rode by gently shifting her weight. After Hartel retired from competitive riding, she opened the first Therapeutic Riding Center in Europe. Both Hartel and Jubilee inspired the world.

AUGUST 4

Miracles with Paws

> Dogs are miracles with paws.
> —SUSAN ARIEL RAINBOW KENNEDY, American author

International Assistance Dog Week begins the first Sunday in August, a celebration created in 2009 to recognize the devoted dogs that help people worldwide cope with disability-related limitations. It is also a week to honor the puppy raisers and trainers of assistance dogs, and a time to highlight the value and abilities of these dogs.

Assistance dogs provide both physical and emotional help. Some guide people with hearing and vision impairment. Specially trained dogs alert their owners in advance of sudden drop in blood pressure, changes in blood sugar level, epileptic seizure, stroke, and heart attack. Dogs assist by retrieving objects out of reach, pulling wheelchairs, opening and closing doors, turning off and on lights, providing balance and support, reminding to take medications, and helping with dressing and undressing. For people with autism and those with post-traumatic stress disorder, assistance dogs provide a calming influence and a sense of security, helping their owners cope with emotional overload. Common breeds trained as assistance dogs include Labrador retrievers, German shepherds, and golden retrievers, but almost any breed can be trained if the dog is calm and intelligent and has good concentration. Assistance dogs allow persons with disabilities to regain independence and confidence, qualities we all cherish. Dogs are, indeed, miracles with paws.

AUGUST 5

You've Got a Friend in Me

In spring 2011, the General Assembly of the United Nations declared July 30 as International Friendship Day. Many countries had already been celebrating their own Friendship Day on the first Sunday in August. Regardless of the date celebrated, Friendship Day is a time to honor those close to us. We aren't the only species that forms friendships. In the absence of pack members or a parent, some animals seek social interactions with the most unlikely animals.

Bubbles and Bella are one example. An African elephant named Bubbles was a 340-pound (154-kg) baby when she was orphaned in 1983, after poachers slaughtered her family for ivory. She was flown to the United States, where Myrtle Beach Safari in South Carolina adopted her. In 2007, a contractor who was hired to build Bubbles a pool abandoned his female black Labrador puppy, Bella, at the preserve. Bubbles and Bella, orphaned and abandoned respectively, are now inseparable. They walk the grounds together and play with each other in the water. Bubbles throws a ball with her trunk, and Bella, perched on the elephant's back or head, leaps into the water and swims to retrieve it.

The world is full of interspecific "friendships"—giraffe and ostrich, baby hippo and giant tortoise, lion cub and skunk, chimpanzee and white tiger cubs, dog and fawn, dog and pig, cat and rat, and cat and goat. Friendship Day is a good time to acknowledge the emotional bond that other animals form, as we celebrate our own friends—human or otherwise.

The Eagle

> He clasps the crag with crooked hands;
> Close to the sun in lonely lands,
> Ring'd with the azure world, he stands.
> The wrinkled sea beneath him crawls;
> He watches from his mountain walls,
> And like a thunderbolt he falls.
> —ALFRED, LORD TENNYSON, "The Eagle"

Alfred, Lord Tennyson, one of Britain's best-loved poets of the Victorian age, was born on August 6, 1809. Tennyson was a master at describing nature in vivid language. We can easily imagine ourselves in the wings and talons of Tennyson's majestic eagle—the power, the energy, and the vision.

Tennyson chose his subject well to convey appreciation of nature's beauty, a central theme of poetry during the Romantic Era (late 1700s to about 1850), for eagles have long been admired and revered. The eagle was the sacred animal for Jupiter and Zeus, Roman and Greek gods, respectively, of the sky and thunder and also king of the gods. Eagles are inspiring in both appearance and behavior, explaining why, for cultures worldwide, the eagle symbolizes courage and strength. Eagles are among the largest birds of prey and are fearless predators; their keen eyes spot prey a mile (1.6 km) beneath them; they are some of the highest flying birds, soaring up to 15,000 feet (4570 m) above the ground at about 65 mph (105 km/h); they glide for hours on warm updrafts of air; and they swoop down at 200 mph (322 km/h), falling like a thunderbolt.

Gaia

Imagine that all the living and nonliving components of our planet work together as a self-regulating system to maintain optimal conditions for life. Just as organisms adapt to their environments, so they also adapt their environments to themselves. This, in a nutshell, is the Gaia hypothesis, developed in the late 1960s by British chemist James Lovelock while working at NASA's Jet Propulsion Labs, and further co-developed by American microbiologist Lynn Margulis in the 1970s.

This idea of Earth as an integrated whole, a single living entity, is not new. The ancient Greeks believed that Gaia, their goddess personifying Earth ("Mother Nature"), kept the living and nonliving components interacting appropriately to maintain balance and harmony. In the eighteenth century, Scottish geologist James Hutton anticipated the Gaia hypothesis by suggesting that geological and biological processes are interlinked, and that Earth is a kind of super-organism. The Prussian naturalist and explorer Alexander von Humboldt (1769–1859) recognized life, climate, and the Earth's crust as one interacting living web. In the twentieth century, Ukrainian geochemist Vladimir Vernadsky maintained that living organisms shape and reshape Earth just as physical forces do.

For those who support the Gaia hypothesis, today is Gaia Consciousness Day—a day to honor Earth and to acknowledge that she is a living, breathing entity. For all of us, it could be a day to become more planet-centered and less human-centered—a day to acknowledge the harm we cause Earth through pollution, deforestation, human-caused climate change, and other depredations, and then amend our ways.

AUGUST 8

Hanzaki Festival

Once thought to be a mermaid or a monkey-fish, the nearly 5-foot (1.5-m) Japanese giant salamander is now a national treasure of Japan. The animal is valued not only because it is unique to the country, but also because of its special appearance—wrinkled skin covered with folds and tubercles, tiny eyes, and a broad flat head. The animal's shape and size give it the air of a large amphibian that roamed Earth long before the dinosaurs. One local name for this salamander is *hanzaki*, meaning "cut in half," referring to the (incorrect) belief that it can regenerate if bisected.

The *hanzaki* has great cultural significance for the Japanese, dating back to a seventeenth-century tale in which a 10-foot (3-m) giant salamander marauded across the countryside eating horses and cows. A samurai named Mitsui Hikoshiro allowed himself to be swallowed whole by the salamander. Mitsui slew the creature with his sword and crawled back out through the stomach. When crops failed and people (including Mitsui) died mysteriously, the salamander's spirit was blamed. To appease the spirit, the people of Maniwa City (Okayama Prefecture, Honshu island) built a wooden shrine to honor the *hanzaki*. This shrine still stands, and the giant

salamander is still honored. Every year on August 8, costumed residents of Maniwa City parade in the streets doing the *hanzaki* dance, pulling two floats, each with a 30-foot (9-m) replica of a Japanese giant salamander.

Wildlife Spotter

How common are common wombats in New South Wales, Australia? How many northern bettongs ("rat-kangaroos") live in far northern Queensland? How many dingos, emus, and kangaroos roam the central arid zone of the Northern Territory? To help answer these and other questions about threatened species, scientists have placed automated cameras throughout Australia. Movement triggers a photo, day and night. Imagine how many images of wildlife accrue per day, month, and year.

Scientists have enlisted help from citizen scientists to identify the animals in these millions of images. The project, called Wildlife Spotter, is sponsored by ABC Science in collaboration with the Australian Museum, various universities, and other organizations for Australia's National Science Week 2016. The project kicked off on August 1, and is ongoing indefinitely. Volunteers sift through images and identify the animals. A short online tutorial is available should one need help. Five people view each image to increase accuracy. To join, go to www.wildlifespotter.net.au. You can invest 10 minutes, ten hours, or as long as you want. All information gathered will be analyzed and uploaded onto the Atlas of Living Australia site: http://www.ala.org.au. The result will be a gigantic picture of Australian wildlife, from dry rangelands to tropical rainforests. When I logged on to the site on August 9, 2017 (a year into the project), 59,773 wildlife spotters already had identified 3,934,495 animals; 3,183,373 images had been completed. Science gains, and nonscientists become invested in protecting Australia's unique fauna.

AUGUST 10

Noble, Powerful, and Brave

At one time they were among the most widespread large terrestrial animals on Earth. They appear in the Chauvet Cave of southern France, which houses the world's oldest known prehistoric cave paintings. They guard our temples and adorn our flags, coats of arms, coins, paper currency, stamps, and brass doorknockers. They are represented in literature, film, paintings, sculpture, ceramics, and jewelry, symbols of royalty, bravery, strength, authority, and power. They are lions, the "King of Beasts." Humans have long been fascinated with lions, as revealed in the significant cultural role they play in our lives.

Now, in the twenty-first century, lions are declining. Between 1993 and 2014, population sizes have declined an estimated 43 percent. Humans are the greatest threat to lions. We destroy their habitat and kill them in defense of human life and livestock. Demand for their bones for traditional Asian medicine is increasing; drinking crushed lion bone is believed to convey increased strength, bravery, virility, and longevity.

August 10 is World Lion Day, a campaign to celebrate these second-largest of all living cats. Over and above the aesthetic value of lions for people, lions play a critical role as an apex predator in the savanna ecosystem. Lions prey on large herbivores such as zebras, giraffes, and wildebeests. If these herbivore populations go unchecked, both herbivores and omnivores will have less to eat. If we lose lions, the savanna ecosystem will suffer, and we will lose an age-old part of our culture.

AUGUST 11

King Puck

Old Irish folklore advised keeping a goat with a herd of cattle because it ensured cows' fertility and prevented them from calving prematurely. In addition, the goat was believed to see the wind and detect imminent bad weather, after which it led the cattle to shelter. Most prominent of all Irish goat folklore, however, is King Puck.

Irish legend tells that during the mid-1600s, when Oliver Cromwell's soldiers were plundering the island, a billy goat alerted the inhabitants of Killorglin of danger. The goat herded the farm animals and led them to the mountains for safety. To give thanks to the billy goat, the townspeople celebrated the goat in an annual livestock fair called the Puck Fair ("Fair of the He-Goat"), held annually ever since.

The Puck Fair is held in Killorglin, County Kerry, Ireland, from August 10 to 12. Residents of Killorglin capture a wild mountain goat from the nearby mountains. A young schoolgirl selected to be Queen of Puck crowns the goat "King Puck." The goat is then caged

and reigns over the town for three days. The coronation signals the beginning of the festival, with parades, concerts, storytelling, traditional music and dancing, fireworks, horse and cow fair, and craft booths. After the festival, the goat is taken back to the mountains and set free.

AUGUST 12

Love Those Elephants

August 12 is World Elephant Day, launched in 2012 to draw attention to the plight of elephants and to encourage support for their conservation. Populations of both African and Asian elephants are declining due to human encroachment, habitat loss, and poaching.

There are many things to love about elephants. Anyone familiar with Rudyard Kipling's *Just So Stories* knows how elephants acquired their outlandish trunks. It happened one day down by the great, grey-green, greasy Limpopo River. Crocodile grabbed Elephant Child's little button nose and pulled and stretched until his nose was several feet long. It never shrank back to a button nose. Elephants express empathy, grief, joy, and anger toward each other. They form strong family bonds. A calf is raised and protected by the entire herd, led by the matriarch. Elephants are extremely intelligent, with a memory that outshines that of some of us. They have cooperative spirits and exhibit loyalty.

More than 65 wildlife organizations and many individuals who support World Elephant Day are working to protect these giants, but elephants themselves also contribute financial support. Captive elephants in zoos and other facilities have been taught to paint as a way of keeping them mentally stimulated, praised for their efforts and rewarded with treats. Their unique paintings are sold by nonprofit organizations, such as The Asian Elephant Art & Conservation Project, to support elephant conservation. Some of the artwork is exquisite!

Festival of Torches

August 13 in ancient Rome was the beginning of Nemoralia ("Festival of Torches"), a women's festival in honor of Diana, goddess of the hunt, goddess of the moon, and virgin goddess of childbirth and women. Nemoralia was a time of rest for women and slaves, and hunting or killing of any beasts was forbidden from August 13 to 15.

Roman writers Ovid and Plutarch, from the first century CE, left us vivid descriptions of Nemoralia. In preparation for the festivities, Roman women prepared candles by dipping rolled papyrus in melted beeswax. They adorned their freshly washed hair with wreaths of flowers, and then hiked or traveled by chariot to Nemi, about 19 miles (30 km) southeast of Rome, where they celebrated Diana, requested her help, and offered fruit and tiny sculptures of stags in her honor. On the altar, they left small baked clay figures of body parts in need of healing, to beseech Diana for help. Festival-goers wrote prayers to Diana on ribbons and tied them to trees. They strolled to a grove and walked in a procession around Lake Nemi ("Diana's Mirror"), where shimmering reflections of their torches and candles danced on the lake surface.

Nemoralia was such a popular festival that the Catholic Church later adopted it as the Feast of Assumption (celebration of the belief that God took the body of the Virgin Mary into Heaven following her death), to be celebrated on or around August 15.

Made from Spare Parts

August 14 is World Lizard Day, an occasion to educate the public about the biology and ecological value of the more than 6100 species of these reptiles. Lizards are critical components of many food webs, serving both as predators and as prey. They are yet another group of animals that are declining worldwide, largely due to habitat modification. Chameleons are one of the most imperiled groups of lizards; at least 36 percent of species are threatened with extinction, because of habitat destruction and collection for the pet trade.

Chameleons are wonderfully bizarre lizards, with bulging eyes that move independently, allowing them to see in two directions at once. They walk slowly, deliberately. Most species have prehensile tails, which tightly grip branches. They can gulp air and become twice their normal size to impress females or rival males and to avoid predation. Chameleons hunt with ballistic weapons—their tongues. Up to twice the length of a chameleon's body, the tongue shoots out from the mouth and snags prey in less than one-tenth of a second. Males of some species have horns, which they use to ram their opponents while trying to push them off branches. Chameleons can change color in less than 30 seconds in response to their moods, to communicate with each other, and in response to temperature. An African myth captures the essence of these animals: The Devil made chameleons from spare parts. He gave them the tail of a monkey, the skin of a crocodile, the tongue of a toad, the horns of a rhinoceros, and the eyes of a "who-knows-what."

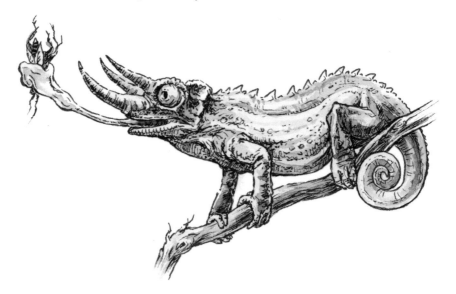

The Enormous, Inflated Bull Frog

Oh the places you will go! The creatures you will see! The stories you will tell, from three thousand feet beneath the ocean surface, where no living human has been.

Explorer, naturalist, ornithologist, and marine biologist, William Beebe began exploring the ocean off the coast of Bermuda with his homemade diving helmet in 1925. Captivated by the richness, colors, and shapes of ocean life, Beebe wanted to dive deeper; to do so, he needed a submersible chamber. In the late 1920s, Beebe and

a wealthy, adventurous engineer named Otis Barton joined forces. Barton provided the blue prints and funding for building a spherical, pressurized vessel, dubbed the bathysphere by Beebe. Barton described his vessel "rather like an enormous inflated and slightly cockeyed bull frog." It was the first submersible chamber that would allow humans to reach the ocean depths.

In June 1930, Beebe and Barton conducted a successful unmanned test of the bathysphere down to 1430 feet (435 m). Between 1930 and 1934, Beebe and Barton made 35 dives of increasing depths off Bermuda—the first humans to observe deep-sea life in the natural environment. On August 15, 1934, the explorers made their record trip, descending 3028 feet (923 m). Amazing sights never before seen by humans, revealed by a searchlight through the bathysphere window, greeted their eyes: phosphorescent fish; schools of jellyfish shining pale green lights; and fish with unbelievably long, sharp teeth.

AUGUST 16

Flooding of the Nile

Without the Nile River, ancient Egyptian civilization as we understand it could not have developed. The Nile, the longest river on Earth, was the primary source of water for the ancient Egyptians in their otherwise arid landscape. Early farmers in the Nile River Valley 5000 years ago depended on the river to grow their crops. Beginning in late June or early July, the Nile overflowed its banks, and by September or October the floodwaters receded and left behind a strip of fertile black soil 6 miles (10 km) wide along each bank. The Egyptian year was divided into three seasons, based on crops and the Nile: Inundation, Growth, and Harvest.

Today is the second day of the Flooding of the Nile Holiday in Egypt, known as Wafaa El-Nil. The holiday begins on August 15 and continues for two weeks. Whereas in ancient days, a beautiful virgin girl purportedly was sacrificed in the Nile to ensure a bountiful harvest, the modern festivity often includes the symbolic throwing of a wooden doll dressed as a bride into the Nile. Today's festivities also include water-skiing, rowing, swimming, poetry readings, concerts, and flower and boat parades. Although the completion of the Aswan High Dam in 1970 ended the annual flooding cycle of the Nile, Egyptians still celebrate this holiday in honor of the river and its agricultural significance over many thousands of years.

AUGUST 17

Going Somewhere

A black cat crossing your path portends misfortune and death. Black cats are demonic, familiars to witches. Many Westerners have grown up with those myths, but black cat folklore varies among cultures. Ancient Egyptians kept black cats in their homes to curry favor with Bast, the cat goddess. In Japan, black cats are appreciated for bringing good luck.

Black cats are genetic mutants. Melanism (unusually high concentration of the dark pigment melanin) is a naturally occurring mutation found in at least 11 species of cats, including leopards, jaguars, and bobcats. Black color may provide a camouflage benefit in dense forest, where light levels are low, allowing cats to ambush prey effectively. But black cats are conspicuous when they venture into open areas, so why isn't this mutation selected against? The gene for melanism may confer another selective advantage: resistance to viral infections.

Today is Black Cat Appreciation Day, a day to dispel black cat superstitions and encourage people to adopt these felines. In the United States, black cats are the least likely cats to be adopted from shelters and therefore experience the highest rate of euthanization. In truth, black cats are no different from any other color of cat. A Chinese proverb states: "Black cat or white cat: If it can catch mice,

it's a good cat." A black cat crossing your path won't bring bad luck. Instead, in comedian Groucho Marx's words: "A black cat crossing your path signifies that the animal is going somewhere."

Pleistocene Re-wilding

Imagine if all large herbivores, including elephants, giraffes, zebras, warthogs, and impalas, disappeared from the Serengeti. It would be an entirely different ecosystem because of the key role these herbivores play in structuring plant communities. Humans likely played a major part in extinguishing giant sloths, horses, giant armadillos, mastodons, and other large herbivores in the New World. Our ecosystems are now very different.

In 2004, Brazilian ecologist Mauro Galetti suggested that elephants, hippopotamuses, horses, and other proxies of extinct large herbivores could be introduced to designated areas he called "Pleistocene Parks" in the Brazilian cerrado (tropical savanna) and pantanal (tropical swampland), as an experiment to better understand the effect of large herbivores on ecosystem structure. On August 18, 2005, American ecologist Josh Donlan and colleagues published an editorial in *Nature* proposing that elephants, Bolson tortoises, camels, lions, and cheetahs be introduced to protected areas in the United States "to restore some of the evolutionary and ecological potential that was lost 13,000 years ago." Critics of the so-called "Pleistocene re-wilding" idea argue that it is unrealistic to assume that today's biological communities function in the same way they did during the late Pleistocene. A spirited debate continues.

The Bolson tortoise, a Pleistocene remnant that once lived in the desert grasslands of Arizona, New Mexico, and Texas, now is found only in northern Mexico. An experiment has begun, though touted as an endangered species recovery project rather than "re-wilding." In 2006, Bolson tortoises were introduced onto two of media mogul Ted Turner's ranches in New Mexico. Biologists are currently monitoring the ecosystem for change.

AUGUST 19

Guide to Heaven

Gladiolus is the birth flower for August, representing moral integrity, sincerity, infatuation, and remembrance. Flower colors include red, pink, pale purple, white, yellow, and orange. The genus name, *Gladiolus*, is the diminutive of the Latin word *gladius*, meaning "sword," in reference to the plant's narrow, sword-shaped leaves. It is sometimes called "sword lily," though the plant is a member of the iris family, not a lily.

Although some species of gladiolus cause skin irritation when handled, the plant has been and still is used for medicinal purposes. During sixteenth-century England, mashed corms (bulblike underground stems) were used as a poultice to extract thorns and splinters. The British also used powdered, dried corms mixed with goats' milk to soothe colic in babies. In parts of southern and western Africa, ground gladiolus corms are used to treat dysentery, diarrhea, stomach upsets, and colds, effective because of the corms' antimicrobial properties.

Watch out to whom you gift a bouquet of gladioli, if that person knows that a symbolic message from the Victorian language of flowers was: "You pierce my heart. I am infatuated with you." Today gladioli are more often given in remembrance or in sympathy for a loss. Showy and sturdy, they represent strength of character. The flowers are often used in elegant spray arrangements at funerals. Chinese folklore tells that gladioli guide the deceased to heaven.

AUGUST 20

For the Love of Honey

The third Saturday in August is National Honey Bee Day in the United States. Humans have exploited bees for millennia. Beekeeping likely began about 9000 years ago, in North Africa, and the first artificial hives were probably pottery urns. In 2015, an estimated 1.7 million tons of honey were produced worldwide. By 2022, the global honey market is projected to reach 2.4 million tons. China is the largest producer of honey, accounting for about 28 percent of the world's total, and is also the world's largest consumer. Honey isn't merely eaten. It is used in medicinal remedies, from acne cure to wound healing, because of its antibacterial properties, and in beauty products including face moisturizers, shampoo, and hair conditioners. Purportedly, honey improves digestion and circulation, and relieves sore throat, eczema, and hangovers.

A (surely apocryphal) tale from Ireland relates that when the Greek philosopher Aristotle wished to discover the secret of how bees make honey, he constructed a glass hive so that he could watch the bees go about their work. When the bees plastered the inside of the hive with wax, preventing him from watching, Aristotle kicked and broke the hive in a rage. Bees stung him until he was temporarily blinded. The story ends with the statement that there were only three things that Aristotle was never able to understand: how bees make honey, the ebbing and flowing of the tides, and the mind of a woman.

AUGUST 21

The Sun Goes Dark

On August 21, 2017, I sat on my back porch ready to watch the solar eclipse that would begin at 10:15 a.m. in Logan, Utah. People within a 70-mile swath of the United States, from northern Oregon to South Carolina, would be in the "Path of Totality" and experience a total eclipse. In northern Utah we would see a 95 percent event.

Throughout history, human cultures have held diverse interpretations of astronomical events. For the ancient Greeks, a solar eclipse signified that the gods were angry. Chinese lore tells that dragons consume the sun; for Vietnamese it is a giant frog, and in parts of South America jaguars swallow the sun. The eclipse ends when the beast regurgitates the hot ball of fire. Sixteenth-century Aztecs believed that after a total solar eclipse, the demons of darkness came to Earth and devoured people. Myths from Germany and western Africa tell that a solar eclipse happens when the sun and moon have sexual intercourse. The Navajo and Eastern Shoshone of North America view a solar eclipse as a time of renewal, both astronomical and personal.

Through my eclipse shades, I watched the moon gradually blacken the sun. By 11:34, the sun and moon crossed paths for maximum coverage from my vantage point, reducing the sun to a golden crescent. I shivered and felt strangely agitated. I understood why a solar eclipse might be interpreted as an evil omen. It was simultaneously magical, surreal, and unsettling. By 12:55 p.m., the sun was again unobstructed, the day was warm and sunny, and I felt calm.

AUGUST 22

Pachamama

In the age of the Inca Empire (ca. 1425–1535 CE), one of the Inca's main deities was Pachamama ("Earth Mother"), a fertility goddess who presided over planting and harvesting. Because she was Earth itself and sustained all life, the Inca frequently honored her to show respect and to give thanks. Llamas were among the Inca's most valuable possessions; they provided wool and meat, and served as beasts of burden. And so, the Inca sacrificed llamas to Pachamama. Before sowing their crops, and then before irrigating their fields, and then again, at the beginning of harvest, they burned a llama and cloth woven from llama wool in her honor.

Many of their descendants, who today live in the Andes of Bolivia, Ecuador, Peru, northern Argentina, and Chile, still honor Pachamama by observing rituals year-round. They bury llama fetuses under the foundations of their homes as an apology to the goddess for cutting into her. Before sipping from a bottle of beer, wine, or chicha (a fermented drink usually made from maize), they pour the first few drops on the ground, for Pachamama. During the harvest month of August, people offer flower petals, coca leaves, chicha, llama fetuses, sacrificed guinea pigs, potatoes, and grains to Pachamama. Depending on the locality, culture, and type of ritual, they either bury or burn these offerings. Either way, they give back to Pachamama in thanks for what they take from her.

AUGUST 23

Vulcanalia

Ancient Romans celebrated Vulcanalia on August 23, a festival to honor the Roman god Vulcan, god of both beneficial and destructive fire. The festival was held during harvest time and the heat of summer, when the land was parched and at risk of burning. Romans prayed to Vulcan to protect their granaries and fields from fire. The god also was seen as a powerful fertility symbol, and so people appealed to him for productive fields. On this day, Romans began their work by candlelight, to curry favor with Vulcan and beseech him to use fire beneficially. Later in the day they lit bonfires and sacrificed fish to the god. Symbolically, the fish, instead of people or grain, were to be consumed by fire.

In his destructive role, Vulcan was god of volcano fire. On August 24, 79 CE, just one day after the Vulcanalia festival, Mount Vesuvius (on the west coast of Italy overlooking the Bay and current city of Naples) rumbled, spit clouds of rock, ash, and fumes, and then spewed molten rock and pulverized pumice at an estimated rate of 1.5 million tons per second. The eruption destroyed the cities of Pompeii and Herculaneum, killing an estimated 13,000 people. Survivors may have wondered if their offerings the previous day had displeased Vulcan.

Today, Mount Vesuvius is the only active volcano in mainland Europe. It last erupted in 1944. Vesuvius is dormant at the moment, but Italian officials have a plan to evacuate nearly 700,000 people from Naples and surrounding towns by buses and trains if the volcano threatens to explode. Fate will not be left to Vulcan.

Mother of Nature Education

Anna Botsford Comstock began to love nature as a child growing up on a farm in western New York. She graduated with a degree in natural history from Cornell University in 1885 and became Cornell's first female professor in 1897. Comstock died at her home in Ithaca, New York, on August 24, 1930, a leader of the Nature Study movement.

The Nature Study movement of the late nineteenth and early twentieth centuries came at a time when people were concerned about land and resource conservation for future generations. Comstock and others advocated observing nature first-hand rather than merely learning from books. She was one of the first to take students and teachers *outside* to observe nature. In 1895, Comstock designed an experimental course of nature study for elementary schools in Westchester County, New York. The program was later adopted statewide. In 1897, she wrote and illustrated *The Handbook of Nature Study*, published in 1911. The book was used by elementary school teachers for decades, has been translated into eight languages, and is still in print.

Fifty-eight years after her death, Comstock was inducted into the National Wildlife Federation Hall of Fame. Referred to as the "Mother of Nature Education," Comstock has motivated generations of children and teachers to connect with nature by direct observation and inquiry. The aim of her approach to nature study was to "cultivate the child's imagination, love of the beautiful, and sense of companionship with life out-of-doors." Her goal is still central to nature education.

AUGUST 25

America's Best Idea

> National parks are the best idea we ever had. Absolutely American, absolutely democratic, they reflect us at our best rather than our worst.
> —WALLACE STEGNER, writer and environmentalist

The idea behind national parks is protection of splendid scenery for no other reason than so people can admire the beauty of the landscape itself. The first US national park, Yellowstone, located in Wyoming, Montana, and Idaho, was established in March 1872. Then came Yosemite and Sequoia, in California, in 1890. By 1916, there were 35 national parks and monuments. On August 25, 1916, President Woodrow Wilson signed the Organic Act of 1916, creating the National Park Service as a new federal bureau in the Department of the Interior, to protect the existing national parks and monuments and any future such areas that might be established. The act stated that the mission of the National Park Service was "to conserve the scenery and the natural and historic objects and the wild life therein and to provide for the enjoyment of the same in such manner and by such means as will leave them unimpaired for the enjoyment of future generations." One hundred years after the signing of the Organic Act of 1916, over 330 million people visited US national parks in 2016. As of August 2017, there are 59 US national parks and 129 national monuments.

The concept of a national park system has since been embraced worldwide. The United Nations Environment Programme estimates that protected areas cover 15.4 percent of the world's land area.

AUGUST 26

No Sloppy Kisses Withheld

Dogs have adapted to living with us for at least the past 16,000 years, although recent research suggests that we might have coexisted twice that long. August 26 is National Dog Day, established in 2004 as a day to celebrate dogs for all they give to us.

A good way to honor dogs might be to incorporate their positive qualities into our lives. For example, *Love without conditions*. Your dog loves you unconditionally, and is always ready to comfort you. We could be more like dogs in our interactions with loved ones. *Live for the moment*. Your dog doesn't obsess about the disappointments of yesterday or worry about tomorrow's happenings. All that matters is the present. We would reduce stress if we lived more in the here and now. *Forgive*. If you return home late, forget the bedtime treat, or skip the daily walk, your dog forgives your misbehavior. No sloppy kisses withheld; no grudges held. We could be more forgiving. *Enjoy the journey.* It's not the destination that's important to your dog. It's the walk itself—the smells and the sights. Being with you. We need to focus more on the journey.

The 2016–2017 National Pet Owners Survey, conducted by the American Pet Products Association, estimates that 60 percent of US households have at least one dog. Imagine how many people worldwide own dogs. Just think what a different place the world would be if all dog owners embraced the canine qualities of love unconditionally, live in the present, forgive, and enjoy the journey. We just might become as terrific as our dogs already think we are.

AUGUST 27

Hoping to Catch a Fly

> It was nestled between two rocks, on an otherwise arid and seemingly death-filled spit of gray and dusty land. This was the very first living thing that could with certainty be said to have appeared after the catastrophe: and it was, Cotteau wrote with measured excitement, *a microscopic spider*. He looked hard for another, but could find only this single specimen. Yet significantly, and with a nice symbolism, "This strange pioneer of the renovation was busy spinning its web!" The lonely little spider was hoping, in other words, that it would eventually be lucky enough to catch a fly.
>
> —SIMON WINCHESTER, *Krakatoa: The Day the World Exploded*

Krakatoa, located between Java and Sumatra in Indonesia, began erupting in May 1883, died down, and then erupted again in June. By August 25 eruptions intensified, and on August 26 Krakatoa spewed heavy ash and hot pumice. On August 27 Krakatoa exploded at 10:02 am. Over 36,000 people lost their lives from the eruption

and associated tsunamis. Krakatoa released an estimated 20 million tons of sulfur into the atmosphere. The island of Krakatoa had almost disappeared, and most of the surrounding archipelago had been destroyed. By August 28, the volcano was silent.

Belgian biologist Edmond Cotteau, a member of a French government-sponsored expedition to explore the aftermath of the Krakatoa eruption, spotted the first stirrings of life. Cotteau found his spider in May 1884, nine months after Krakatoa exploded. It represented hope for a new beginning.

AUGUST 28

For the Birds

Some kids called him "Professor Nuts Peterson." He'd carry a snake in his pocket or a bird's egg in his cap. At eleven he was already in junior high school and yet refused to walk in line with the other kids. He looked thin and gawky as a fledgling egret and sometimes smelled of skunk.

—PEGGY THOMAS, *For the Birds: The Life of Roger Tory Peterson*

Roger Tory Peterson, naturalist, ornithologist, and artist, was born in Jamestown, New York, on August 28, 1908. As a child, Peterson pressed wildflowers, netted butterflies, collected abandoned bird nests, watched birds, and sketched birds in the margins of his schoolbooks.

Peterson published his first book, *A Field Guide to the Birds*, in 1934. The book cost $2.75 and sold out the first printing of 2000 copies within one week of publication, making Peterson famous almost overnight. All over the United States, people began bird-watching, now able to identify birds with their "Peterson field guide." After WWII, Peterson revised his field guide and wrote a guide to the birds of Europe. More than seven million copies of Peterson's *A Field Guide to the Birds* (6 editions) have been sold. Today, more than 50 books in the Peterson Field Guide Series encourage us to care about nature and help us identify animals, birds' nests, animal tracks, plants, mushrooms, rocks, stars, and planets.

AUGUST 29

The Letter

On August 29, 1831, 22-year-old Charles Darwin returned home after a geology fieldtrip in Wales to find a letter from his friend and mentor John Stevens Henslow, clergyman and professor of botany at Cambridge University. The letter would set the course for Darwin's life and change biology forever. Henslow had been offered the post of naturalist and "gentleman companion" to join Captain Robert FitzRoy on a two-year trip to survey South America on the *HMS Beagle*. Henslow's wife dissuaded him from accepting the post, and it was then offered to Henslow's brother-in-law, who accepted but then changed his mind. At that point, Henslow recommended his protégé and favorite pupil, Charles Darwin. Darwin persuaded his father to allow him the opportunity, and four months later Darwin set out on what would be a nearly five-year voyage during which he studied geological formations and made natural history observations and collections.

What if Darwin's father had not allowed his young son to take the post? Known to be rather stern, Robert Darwin initially felt it would be a waste of his son's time. Would Charles Darwin have pursued his interest in natural history in some other direction? We each leave our mark on the world, hopefully making it a better place. For many of us, there is one defining event that determines our path to leaving that mark. For Charles Darwin, the letter he received on August 29, 1831, paved the way for his theory of evolution via natural selection, the basis for all of biology.

AUGUST 30

Eradication by Mastication

According to the Institute for Applied Ecology (IAE), invasive species cost the world over $1.4 trillion each year. The annual cost of the impact, including lost economic productivity and control efforts to remove invasive species, is estimated at 5 percent of the world's economy. Invasive species cause declines and local extinctions of native species through competition, predation, introduction of disease and parasites, and habitat alteration.

The annual Invasive Species Cook-off, also known as "Eradication by Mastication," is a fundraiser for IAE held in late August in Corvallis, Oregon, that raises public awareness of the problem. Join the potluck, and if you think you've made a great gastronomic creation, enter it in the competition along with offerings by local chefs. There's no end to the possibilities: Himalayan blackberry pie, garlic mustard pesto sauce or ice cream, Japanese knotweed jam, fried bullfrog legs, grilled wild boar chops, barbequed nutria nuggets, steamed purple varnish clams, baked Asian carp, boiled red swamp crayfish, and European green crab cakes.

You could start an annual tradition in your hometown to fight against your local invasive species by organizing an "Eat the Invaders Banquet." Try making dandelion spanakopita, starling pie, kudzu quiche, or Asian carp kebab. Include live music for atmosphere and pamphlets to educate your participants on the evils of invasive species. *Bon appetit*!

From Sacred Ibis to Mascot

The ibis, a large, long-legged wading bird with a distinctive, down-curved bill, holds a special place in Egyptian mythology. Ancient Egyptians revered Thoth, god of knowledge, wisdom, the moon, and magic, and patron of scribes, writing, and science. Thoth most often was depicted as a man with the head of an ibis. The bird itself represented Thoth, and to honor the god, priests raised and sacrificed sacred ibises. Archeologists have found millions of mummified ibis remains in Egyptian catacombs and tombs. Some were buried with snails in their bills, while others had been stuffed with fish and grain, suggesting that food was offered for the birds' afterlives. From a practical standpoint, ancient Egyptians valued ibis because they improved fishponds by eating snails that harbored dangerous liver parasites. Ancient Egyptians celebrated the Festival of Thoth during August, the season of flood. The ibis was considered the herald of the flooding of the Nile, a critical event for farmers living along the river.

Fast forward in time to current popular culture. North American folklore holds that the ibis is the last to seek shelter before a hurricane, and the first to emerge after the storm ends. It is a bird of courage and leadership in the face of danger. Reflecting that lore, Sebastian, a tough-looking American white ibis, serves as the mascot of the University of Miami—"the Hurricanes."

Martha

In 1805 I saw schooners loaded in bulk with Pigeons caught up the Hudson river, coming in to the wharf at New York, when the birds sold for a cent a piece. I knew a man in Pennsylvania, who caught and killed upwards of 500 dozens in a clap-net in one day, sweeping sometimes twenty dozens or more at a single haul. In the month of March 1830, they were so abundant in the markets of New York, that piles of them met the eye in every direction.

—JOHN JAMES AUDUBON, *The Birds of America*

John James Audubon also wrote that passenger pigeons would decline only by "gradual diminution of our forests," not by human exploitation. Sadly, he was mistaken. During the early nineteenth century, passenger pigeons were the most abundant birds in North America, with an estimated total number approaching 5 billion. The flocks that numbered in the hundreds of millions in 1800 declined gradually until about 1870. Then, over the next two decades the number of birds in flocks plummeted to mere dozens. They were slaughtered and sold as cheap food and their feathers used to adorn hats, until the last wild bird was shot in 1900 or 1901. The passenger pigeon had been hunted out of existence. Martha, the last known passenger pigeon, died at the Cincinnati Zoo on September 1, 1914.

Passenger pigeons were victims of the fallacy that no abundant species could ever be exploited to extinction. The tragedy served as a wake-up call that we cannot take the survival of any species for granted.

Animal Remembrance

> Until one has loved an animal a part of one's soul remains
> unawakened.
> —ANATOLE FRANCE, French poet and novelist (1844–1924)

September is World Animal Remembrance Month, sponsored by humane societies, animal rescue centers, veterinary clinics and hospitals, and other animal-oriented organizations and businesses. The sponsors promote it as a time to appreciate service canines that have improved human lives—guide dogs, medical alert dogs, therapy dogs, and mobility assistance dogs; a time to reflect on wild and domesticated animals that have perished in natural disasters; a time to honor animals that have served us during wartime—dogs, horses, dolphins, and pigeons. It is a time to mourn victims of human activity: the estimated 25 million dogs eaten each year in South Korea, the Philippines, China, Vietnam, and elsewhere; the millions of animals that suffer and die during "blood sports," including dog fighting, ram fighting, cock fighting, betta (fish) fighting, and cricket and spider fighting; and the bulls killed by matadors for spectator thrill amid pomp and pageantry.

On a personal level, World Animal Remembrance Month is a time to honor the memory of pets we have loved and that have loved us back—especially the ones that have awakened our souls. Undoubtedly, these pets heightened our empathy and encouraged us to become better people.

Reality of Death

In late August or early September, following ancient tradition, Hindu families in Nepal who have lost a relative in the past year form a procession through the city streets, each leading a cow. (If a cow is not available, a child dressed in a cow costume is substituted.) As the most revered of all animals for Hindus, cows are believed to assist a deceased's journey to heaven. This annual procession is called Gai Jatra, or the festival of cows. The festival honors the recently deceased, but just as importantly, it is a time for the living to see that they are not alone in grieving loved ones. The festival aims to help people accept the reality of death and to prepare for that reality themselves.

Customs associated with the festival vary from city to city. The main event, and one of the most joyous, is held in Kathmandu. There, as the cow procession meanders along streets, people watching the procession offer small packets of fruit, sweets, and oats to the families to sustain them during their walk. After the procession, everyone dresses in costume, wears a mask, and sings late into the evening. At least for a while, people set aside their sorrow, enjoy being alive, and reflect that death is a natural, and inevitable, part of life.

The Disobedient Rabbit

> Once upon a time there were four little Rabbits and their
> names were Flopsy, Mopsy, Cotton-Tail, and Peter. They lived
> with their Mother in a sand-bank, underneath the root of a
> very big fir-tree.
> —BEATRIX POTTER, *The Tale of Peter Rabbit*

Beatrix Potter, English artist and writer, found her inspiration in
nature, and she loved animals and children. During the 1890s, she
sent illustrated stories in letters to the children of her former governess, Annie Moore. One of these stories, written in a letter on September 4, 1893, was about a naughty little rabbit named Peter. The
Moore children kept their letters, and in 1900 the former governess
suggested to Potter that she should publish the stories. Potter borrowed back the letters and the rest is literary history.

Potter sent her Peter Rabbit manuscript and illustrations to six
publishers and received six rejections. She privately printed 250
copies of her book in September 1901 and sent them to family and
friends for Christmas. Then she printed another 200 copies. In October 1902, Frederick Warne & Co. (one of her original rejecters)
published her book. Children worldwide have bonded with this dis-
obedient little rabbit who snuck into Mr. McGregor's garden, lost
his shoes and little blue jacket with brass buttons, and nearly lost
his life.

True to her love of nature, Potter spent her later years protecting
the landscape in England's Lake District. She purchased land with
the proceeds of her children's books, and then donated the land to
the National Trust to be protected for future generations.

SEPTEMBER 5

Snakes as Messengers

Various cultures view snakes as links between their world and their gods' world, an association that may reflect snakes' mysterious movements. Snakes appear seemingly out of nowhere and disappear again; they zigzag in and out of crevices; and some are excellent climbers. The Hopi of the high desert of northern Arizona view snakes in this way.

The people plant corn, melons, gourds, and beans in the mouths of arroyos and in the dry washes that sometimes flood. They irrigate their crops by hand, and they enlist snakes as messengers to carry their prayers for rain to the gods. Every other year from late August to mid-September, the Hopi hold a 16-day celebration known as the Hopi Snake Dance, culminating in the dance itself. Preparations for the dance take place during the last nine days of the celebration. Snake-priests catch rattlesnakes, bull snakes, and desert striped

racers. The snake-priests bless the snakes, ceremoniously wash them in water infused with herbs, and exchange spirits with them. In the final ceremony, the snake-priests dance while holding the live snakes in their mouths. Following the dance, women scatter cornmeal on the ground. The snakes are dropped onto the cornmeal, which covers them as they move about. The Hopi then release the snakes below their mesa homes, where they will carry prayers for rain to the gods who live in the underworld, along with the spirits of the Hopis' ancestors.

Aristotelian Celebration

The ancient Greek philosopher Aristotle (384–322 BCE) wrote a text called the *History of Animals*, esteemed as the primary source of zoological knowledge for 2000 years. Many of his descriptions of animals, their morphology and their behavior, are correct. But, like all scientists, Aristotle got a few things wrong. Normally an astute observer, he erroneously believed that men have more teeth than women. He wrote that women are "immature" and "deficient," and that men have a more important role in reproduction. Aristotle wrote that snakes gain immortality through shedding their skin, and that life forms were designed with a purpose. Some of his explanations of anatomy are reminiscent of desperate answers offered on today's comparative anatomy exams—for example that the snake's forked tongue provides it with double the gustatory pleasure. Aristotle wrote that some animals come from parents of the same kind; others, such as insects and eels, are generated spontaneously from putrefying mud and vegetable matter. We know differently now, though it wasn't until 1668 that Francesco Redi's experiments with maggots refuted spontaneous generation.

In early September, the University of Latvia, in Riga, welcomes new students with an Aristotle Celebration, an ongoing tradition for over 45 years. Never mind that Aristotle made a few mistakes regarding animals. He is an inspirational role model, as he thought deeply about the world and had a passion for learning.

"Little Toy Dogs"

In May 1804, Meriwether Lewis and William Clark set out from near St. Louis on their Corps of Discovery Expedition to explore and map newly acquired territory of what is now the United States. A secondary objective was to document the natural history of the west. In their journals, the men recorded observations of more than 300 plants and animals, many previously unknown to science. One of these was the prairie dog.

On September 7, 1804, in what is now Boyd County, Nebraska, the expedition encountered a "village" of small squirrel-like animals that French trappers and traders called *petits chiens* ("little dogs"). What a sight it must have been—hundreds of little rodents barking warning calls to each other at burrow entrances and disappearing underground when the men approached. Lewis described the high-pitched whistles as the barks of "little toy dogs."

Members of the expedition caught one for a mascot. It thrived on prairie grasses and rode along to the winter camp at Fort Mandan, North Dakota. Lewis and Clark decided their mascot would make a great gift to President Jefferson. In April 1805, a Corporal and his crew headed down the Missouri River and back east to deliver the gift. The President was delighted. He kept it as a pet for a short while, and in October 1805 gave it to painter and naturalist Charles Willson Peale to display in his museum in Philadelphia, the nation's first natural history museum. There, the live prairie dog stole the show amid the stuffed birds and mammals, Mandan buffalo robe, and other "curious and unusual" animals and artifacts sent back by the Discovery Expedition.

SEPTEMBER 8

Ivan

He begs to have his head and jowls stroked. He adores dandelion greens and cilantro. He watches everything going on. He has the run of the house—under supervision, for he is a mischief-magnet. He is toilet-trained and performs on cue every morning in a pan of warm water. He is Ivan, an 11-year-old, 4-foot-long (1.2 m) pet green iguana.

In 1998, September 8 was established as National Iguana Awareness Day to educate the public about green iguanas and to discourage viewing them as "disposable pets." Green iguanas are one of the most common reptiles in the pet trade. Nearly 1 million baby green iguanas are imported to the United States each year, mostly captive-bred in their native countries from southern Mexico to southern Brazil and Paraguay, and the Caribbean Islands. They are inexpensive, cute, and beautiful. The perfect pet, right? Wrong. Green iguanas grow quickly and have demanding dietary, space, and lighting needs. An estimated 70 percent of captive baby iguanas die within their first year because their needs are not met.

Green iguanas aren't cuddly. They don't purr or give sloppy kisses. So why are they such popular pets? The iguana's draw may be the novelty of watching an exotic reptile up close and personal. Ivan is a valued member of the household, providing companionship and endless entertainment, but his owners have devoted abundant resources to keeping him healthy. The supporters of National Iguana Awareness Day want people to have the facts, should they decide to adopt a green iguana.

SEPTEMBER 9

Experiencing Nature Firsthand

> A nobler want of man is served by nature, namely, the love of Beauty.
> —RALPH WALDO EMERSON, "Nature" (chapter 3: Beauty)

On September 9, 1836, Ralph Waldo Emerson published his essay "Nature," an analysis of why humans do not fully accept the beauty of nature. The essay reflected the philosophical and social movement known as transcendentalism, which includes the view that God pervades all of nature and humanity, that the individual is endowed with intuition and insight, and that we can understand reality by studying nature. Emerson wrote that to truly engage with nature, we must distance ourselves from the distractions of society, go out into nature, and observe her in solitude.

In the introduction of his lengthy essay, Emerson lamented our tendency to accept past-accumulated knowledge and traditions instead of experiencing nature firsthand in the present. Each of the eight chapters—Nature, Commodity, Beauty, Language, Discipline, Idealism, Spirit, and Prospects—offers a unique perspective of our relationship with nature. Emerson ended his essay by emphasizing that whereas we now apply only rational understanding to nature, we would be better served to restore spirituality in viewing nature to rekindle our wonder of the natural world. Emerson's essay has influenced generations of people worldwide, including transcendentalist Henry David Thoreau, who observed nature in solitude for two years, two months, and two days while living in a one-room cabin he built beside Walden Pond, on the property of his friend and mentor, Ralph Waldo Emerson.

SEPTEMBER 10

Sharing the Passion

> Science is an integral part of culture. It's not this foreign thing, done by an arcane priesthood. It's one of the glories of the human intellectual tradition.
>
> —STEPHEN JAY GOULD, *Independent* (London, January 24, 1990)

Biologist Stephen Jay Gould (1941–2002) wrote numerous books for a general audience, including *The Panda's Thumb*; *The Flamingo's Smile*; *Bully for Brontosaurus*; *Hen's Teeth and Horse's Toes*; and *An Urchin in the Storm*. Many of the essays from these books were originally published in Gould's regular column in *Natural History* magazine, where he discussed topics as divergent as natural history, health and longevity, racial stereotyping, evolution and creationism, and the human genome.

Gould was born on September 10, 1941, in Queens, New York City. When he was five years old, his father introduced him to *Tyrannosaurus rex* in the Hall of Dinosaurs at the American Museum of Natural History. Gould was spellbound. The bones may have planted a seed that inspired him to become a paleontologist and evolutionary biologist. Gould spent his career working at Harvard University and the American Museum of Natural History. He received numerous awards for his scientific research, but equally valuable was his communication of science to the public through his writing and speaking engagements.

As scientists, we are trained to be objective. In our objectivity, sometimes we forget *why* we became scientists—our innate curiosity and sense of wonder, a drive to understand the world around us, and the joy of discovery. Gould shared his passion for science with the nonscientist.

SEPTEMBER 11

Who Gets Saved?

In 2012, more than 8000 scientists identified 100 of the most threatened fungi, plants, and animals on Earth—a manageable list of species on which to focus conservation efforts. All of these species are seriously declining, but another criterion for inclusion on the list was that the species have few defenders, because we don't see them as benefiting us in any way. For example, included on the list are the Singapore freshwater crab, angel shark, Tarzan's chameleon, Okinawa spiny rat, Attenborough's pitcher plant, and Chilenito (a cactus). On September 11, 2012, this list was presented at the International Union for Conservation of Nature and Natural Resources (IUCN) meeting in South Korea. The authors emphasized that most of these

100 species can be saved *if* conservation efforts they recommended are carried out.

But will we do it? How do we choose where our effort and conservation dollars should go? Charismatic species, for example sea turtles and giant pandas, and those of direct benefit to us, like trout and salmon, receive the most attention. Public support influences conservation decisions, and unfortunately many people feel that our resources should be earmarked for birds and mammals, rather than for "second-class citizens" like slugs, centipedes, crickets, spiders, sharks, and snakes. We need to encourage a different mindset—that all organisms have worth, not just the charismatic or useful. That change in perception will come only with public education.

SEPTEMBER 12

Created from Fire

King Francis I of France, patron of the arts and initiator of the French Renaissance, was born on September 12, 1494. Of all the personal emblems he could have chosen, this king chose the salamander surrounded by flames. His Latin motto, translated into English, was "I nourish the good and extinguish the bad."

Salamanders have long been associated with fire. Ancient European legend claimed that salamanders were created from fire and could withstand infinite heat, an association that likely arose from people watching European fire salamanders. These yellow-orange and black salamanders take refuge in logs and then appear when those logs are tossed into fireplaces and wood stoves. The Greek philosopher Aristotle (384–322 BCE) wrote that salamanders could not only withstand fires, but also extinguish them. In the first century CE, the Roman scholar Pliny the Elder reinforced this belief by writing that salamanders are so cold that they can extinguish fire by contact. The Italian polymath Leonardo da Vinci (1452–1519) claimed that the salamander has no digestive organs and consumes only fire.

The belief that salamanders could survive fire led to the idea that they could pass through fire "without stain." Christian writers then used the concept as an allegory for the righteous "quenching the flames of lust." Thus, the salamander came to symbolize enduring faith, chastity, purity, virginity, and self-restraint. King Francis chose his emblem well, a symbol of goodness and power.

SEPTEMBER 13

Eagles and Flying Robots

The eagle flies high, swoops, grabs the small drone in his talons, returns it safely to the ground, and claims his reward—a chunk of raw meat. This scenario is the brainchild of Sjoerd Hoogendoorn, a Dutch security consultant who happened to visit New York City during the week of September 11, 2001. Deeply affected by the terrorism, Hoogendoorn developed this idea in hopes of making the world a little safer. On September 13, 2016, the Dutch National Police announced plans to train eagles to take down drones—a world-first for law enforcement. The Dutch efforts may be just the beginning of a new way to retrieve enemy drones.

Drones (unmanned aerial vehicles) have many positive applications, including forest fire detection, pollution monitoring, search and rescue, aerial photography, and counter-terrorism operations. But in the hands of the terrorists themselves, drones can be deadly weapons, loaded with explosives or biological/chemical hazards and crashed into vulnerable targets. The advantage of birds of prey capturing drones is that the bird brings the drone safely to earth. Other methods, such as jamming drone signals or shooting a drone out of the sky with buckshot, cause the drone to crash-land, potentially endangering life and structures below.

Depending on the kind of eagle trained, the bird snatches the drone because it sees the moving object either as food or as another bird of prey encroaching on its territory. Thanks to their amazing

visual acuity, the eagles grab the centers of the drones and avoid the rotating blades. Hunting drones with eagles takes falconry to a new level!

Wonder of the World

Alexander von Humboldt was born into an aristocratic Prussian family on September 14, 1769. As a young man, he developed a passion for plants. His surroundings weren't very diverse botanically, however, and Humboldt dreamed of exploring exotic lands. From 1799 to 1804, he explored Central and South America, covering more than 6000 miles (9650 km) by foot, horseback, and canoe.

Humboldt is recognized as the first scientist to describe the biological and geological wonders of South America. In addition to botany and natural history, Humboldt was deeply knowledgeable about mining and geology, biogeography, and philosophy, among other fields—a veritable Renaissance man. Befitting the most celebrated scientist of his time, Humboldt's name is known throughout the world, bestowed on hundreds of plants and animals, from the Humboldt lily to the Humboldt penguin, and on geographical features, from the Humboldt Glacier in Greenland to the Humboldt Current that flows along the west coast of South America.

Humboldt's popular writings have inspired many other naturalists and scientists. Charles Darwin claimed that Humboldt's book *Personal Narrative of Travels to the Equinoctial Regions of America* was the reason he was eager to join the *HMS Beagle* as naturalist. Henry David Thoreau modeled himself after Humboldt, as he developed a lifestyle in which he could immerse himself in nature. John Muir grew up reading Humboldt's books, which greatly influenced his thinking about ecology and preservation of nature. Ralph Waldo Emerson referred to Humboldt as "one of those wonders of the world."

Antediluvian Animals

Charles Darwin and the *HMS Beagle* landed on the Galápagos Islands on September 15, 1835. Several days later, Darwin wrote in his journal, "In my walk I met two large tortoises, each of which must have weighed at least 200 pounds. One was eating a piece of cactus, and when I approached, it looked at me, and then quietly walked away; the other gave a deep hiss and drew in its head. These huge reptiles, surrounded by the black lava, the leafless shrubs, and large cacti, appeared to my fancy like some antediluvian animals."

Darwin noted morphological differences in the tortoises on different islands. Later, in England, while ruminating on his observations, he began to suspect that the tortoises had changed as they adapted to contrasting environments on the different islands. The two general shapes of tortoises, domed and saddle-backed, do in fact illustrate how animals can diverge morphologically when they are geographically isolated. Tortoises with domed carapaces are found on the wetter, higher islands, where food is abundant and accessible close to the ground. Those with saddle-shaped carapaces occur on arid islands, where food is scarce during droughts. The front of the carapace angles upward like a saddle, an adaptation that allows the tortoise to reach up and eat the pads from tree cacti. The idea that species could change over time as they adapt to new environments played a major role for Darwin in developing his theory of evolution by natural selection.

Crocodile God

On September 16, 1975, Papua New Guinea (PNG), home to nearly 700 ethnic groups, attained full independence from Australia.

Papuan New Guineans celebrate their cultural identities on Independence Day with music and dancing, and by acting out folklore in song. One of my favorite creation stories from PNG is the Kikori story that explains how PNG and its 600 or so small offshore islands originated.

According to the Kikori, in the beginning there was only water and Crocodile God. In time, Crocodile God gave birth to First Man and First Woman. Because there was only water in every direction, Crocodile God allowed the couple to live on his back. The people were fertile, and in time they had produced so many descendants that there was no remaining living space on Crocodile God's back. He told the people to leave. The people found land only on great islands made from Crocodile God's dung, but they were content because the land was fertile and the fishing superb. Crocodile God's dung islands became known as Papua New Guinea, the original home of all mankind.

It is not surprising that the Kikori chose a strong, powerful animal like the crocodile as their creator, nor is it surprising that the theme of overpopulation is woven into the Kikori's creation myth. There is a finite amount of habitable land on PNG, and the people have one of the highest fertility rates in the Pacific region. Folklore provides a lens for understanding human cultures.

Love Charm

Aster, the birth flower for September, comes from the Greek word *aster*, meaning "star." According to Greek mythology, gods and goddesses lived on Earth during the Golden Age. Once people became greedy and wicked, the immortals abandoned Earth. The last to leave was Astraea, virgin goddess of innocence. When she rose to the heavens, she became the constellation Virgo. She wept, out of compassion for humanity. Her tears mixed with stardust, and when they fell to Earth they became star-shaped asters.

Another legend explaining the origin of asters comes from the Cherokee of North America. Two warring tribes fought over prime hunting grounds. The only surviving members were two young sisters who fled to the woods. One wore a doeskin dress dyed lavender-blue; the other wore a bright yellow doeskin dress. That night, Herb Woman looked into the future and saw that the sisters would be hunted down. She sprinkled them with a magic brew and covered them with leaves. The next morning two flowers bloomed where the girls had slept—one a lavender-blue aster, the other a goldenrod.

During the Victorian Era (1837–1901), a gift of asters conveyed a message of love and devotion. A single aster was given as a "love charm" in hopes of eliciting affection from the recipient. Today, asters are one of the more popular flowers in floral arrangements. Available in white, pink, red, blue, lilac, purple, and mauve, they still represent love.

SEPTEMBER 18

Touch

Skin accounts for about eight pounds of an adult human's body. Among other functions, including protection and temperature regulation, skin provides us with sensation. Our skin has about 5 million nerve receptors for sensing touch, temperature, vibration, pressure, and pain. September is National Skin Care Awareness Month, established to stress the importance of taking care of our largest organ.

Our sense of touch develops while we are still in the womb. Once in the outside world, touch comforts a baby. Later on, touch provides another dimension for appreciating and understanding nature. Touch allows us to feel the sandpaper lick from a cat's tongue, and a soft kiss from a lover's lips.

Touch is the most important sense for some animals, and many have body extensions for feeling their environment. Insects and crustaceans have tactile antennae (depending on the group, antennae may also sense heat, vibrations, and chemicals). Most mammals have vibrissae—specialized hairs, for example whiskers. And then there's the star-nosed mole, with the world's most sensitive nose. Twenty-two pink, fleshy appendages that ring the mole's snout allow it to touch up to a dozen objects per second as it tunnels underground in search of food—worms, insects, and crustaceans. More than 100,000 nerve fibers run from the animal's star-studded nose to its brain—more than six times the number of touch receptors in the human hand.

SEPTEMBER 19

Inti and Mama Quilla

Many traditional cultures bestow special significance to the sun and the moon. They are sometimes viewed as enemies, but more often are joined in marriage. The Inca celebrated the Spring Equinox (south of the equator) on or about September 19. It was a time of feasting, dancing, and rejoicing to honor the sacred union of Inti, the sun god, and Mama Quilla, the moon goddess. Inti was the ranking Incan deity. Farmers, especially, revered Inti, as they depended on his warmth to nurture the growing crops. Inti was often represented by a human face on a gold disk with rays.

Mama Quilla ("Moon Mother") marked the passage of time. The Inca used the waxing and waning of the moon to calculate monthly cycles, from which they set the dates for their religious festivals. Women felt a special connection with Mama Quilla, as she regulated menstrual cycles. Silver was considered to be Mama Quilla's tears fallen to Earth, and she was often depicted in human form on a silver disk. The Inca feared lunar eclipses, because they believed the shadow on the moon was a wild animal attacking Mama Quilla. If the animal harmed the goddess, the nighttime would plunge into permanent darkness.

In addition to feasting and dancing during the Spring Equinox, the Inca performed ritualized enactments of the union between Inti and Mama Quilla to celebrate fertility and procreation. It was thought to be an especially good time to conceive.

SEPTEMBER 20

Amur Tiger Stamp

Some people dedicate their lives to conservation research, policy, education and public outreach, advocacy, or law. Some volunteer their services or make major donations to conservation. An innovative, easy-on-the-pocketbook way for all of us to support international wildlife conservation was initiated in 2009. To raise awareness of endangered wildlife, and to contribute to their protection, the World Wildlife Fund proposed issuing and selling a semipostal stamp: a postage stamp sold for postal value, plus an added surtax to raise money for a cause. The Multinational Species Conservation Funds Semipostal Stamp Act of 2009 passed Congress with overwhelming bipartisan support.

On September 20, 2011, the United States Postal Service issued a "Save Vanishing Species" semipostal stamp with a depiction of an Amur tiger cub designed by American illustrator Nancy Stahl. All proceeds go to the Multinational Species Conservation Funds, administered by the US Fish and Wildlife Service. Unless the stamp, which costs 60 cents, is reauthorized, it will not be available past the end of 2017. By August 2017, more than 38 million stamps have been sold. Over $4.1 million has been raised for the Multinational Species Conservation Funds to help protect wild populations of endangered species, including tigers, elephants, rhinos, great apes, and sea turtles.

SEPTEMBER 21

Doomsday Vault

Worldwide, more than 1700 gene banks hold seeds of food crops as a safeguard against the plants being wiped out by war or natural disasters, including disease. Many of these banks, however, are themselves vulnerable to depredations of war, natural disasters, poor management, or freezer malfunction. As a safeguard, in 2008, the Svalbard Global Seed Vault was established on the Norwegian island of Spitsbergen in the remote Arctic Svalbard archipelago, built underground in a permafrost zone about 620 miles (1000 km) from the North Pole. Often referred to as a "doomsday" vault, it serves as a backup for the other 1700 gene banks. Each contributing country owns and controls access to the seeds it deposits, and only it can withdraw the seeds. As of February 2017, the vault has more than 940,000 samples, contributed by most nations in the world. The vault has a capacity of 4.5 million samples, each sample consisting of about 500 seeds.

On September 21, 2015, Reuters announced that the war in Syria had spurred the first request for withdrawal of seeds from the vault. The seeds were requested by the International Center for Agricultural Research in Dry Areas (ICARDA), because researchers could not access the gene bank in the war-torn city of Aleppo, Syria. In February 2017, ICARDA returned more than 15,000 samples, including potato, rice, wheat, and chickpea, to the Svalbard vault. The Svalbard Global Seed Vault is doing what it was designed to do. It is the ultimate insurance policy for the world's food supply.

SEPTEMBER 22

What Is Life?

Buffalo were disappearing from the northwestern plains of Canada during the mid-nineteenth century. The Canadian government needed the First People's hunting grounds to build a transcontinental railroad; settlers wanted the land for cattle. In an effort to control the land, while still wishing to help the First Peoples survive, Queen Victoria made an offer. In return for financial assistance and help in raising cattle and grain, the First Nations were expected to "cede, release, surrender, and yield up" to the Canadian government all rights, titles, and privileges to their hunting grounds.

Chief Crowfoot of the Siksika First Nation (Blackfoot) knew that nothing would stop the invaders, and that his people must accommodate them, thus losing the buffalo. He signed Treaty Number 7 in southern Alberta on September 22, 1877. His signing led the way for other chiefs to sign. Although not all the promises were honored, and many of his people died from starvation and disease, Chief Crowfoot maintained that overt violence would only worsen matters. He died in 1890, respected as a warrior and peacemaker. Reportedly, his last words were:

A little while and I will be gone from among you. Whither I cannot tell. From Nowhere we came; into Nowhere we go.

What is life? It is the flash of a firefly in the night. It is the breath of a buffalo in the wintertime. It is the little shadow that runs across the grass and loses itself in the sunset.

SEPTEMBER 23

Prized Fur Coats

Profit-hungry hunters and fashion-conscious consumers have long lusted for sea otter pelts, at one time considered the most valuable of all animal skins, prized for their thickness. Unlike most marine mammals, which have blubber to shield them from the icy ocean water, sea otters rely on their fur for warmth. They have the densest fur coat of any mammal. To put thickness into perspective, humans have an average of 700 hairs per square inch on the head. Sea otters have up to a million hairs per square inch. Theirs is a two-layered coat—long outer guard hairs and a dense, fine undercoat.

Sea otters once may have numbered a million individuals, but by the early 1900s the fur trade had reduced their numbers to between

1000 and 2000. Currently there are an estimated 106,000 sea otters worldwide. They have made a partial comeback thanks to protection, beginning in 1911 when the first international treaty to protect wildlife was signed by the United States, Great Britain, Japan, and Russia. The North Pacific Fur Seal Convention, designed to manage commercial harvest of both seals and sea otters in the Pribilof Islands, banned pelagic hunting. An exemption was made for Aleut and Aino who hunted seals using traditional methods and for noncommercial purposes.

The last week of September is Sea Otter Awareness Week. Although sea otters are still protected from being hunted, oil spills from offshore drilling and shipping threaten the survival of vulnerable populations.

SEPTEMBER 24

Bluebird of Happiness

> Somewhere over the rainbow,
> Bluebirds fly
> And the dreams that you dreamed of
> Dreams really do come true.
> —"Over the Rainbow," from the movie *The Wizard of Oz*

Today is National Bluebird of Happiness Day. Blue birds have long been regarded as harbingers of happiness in diverse cultures, from ancient China to current-day First Peoples of North America. The birds symbolize joy, prosperity, good health, and renewal of life. Blue birds associated with this symbolism are found in many parts of the world, including the three species of bluebirds (genus *Sialia*) found in North America: the eastern, western, and mountain bluebirds. Polls worldwide reveal that blue, the color of the sky, the oceans, lakes, and springs, is people's no. 1 favorite color. The birds' striking blue feathers reflect nature herself.

Belgian playwright Maurice Maeterlinck coined the phrase "bluebird of happiness" in his play "L'Oiseau bleu" ("The Blue Bird"), which premiered in September 1908 in Moscow. The story tells of a girl Mytyl and her brother Tyltyl, who, with help from the fairy Bérylune, undertake a magical quest through fantasy kingdoms to find the "blue bird of happiness." The children return home empty-handed and realize that the "blue bird of happiness" is Tyltyl's pet bird, at home all along, and that the forest outside their home and their home itself look lovelier than ever. The message of the story is that although we all search for happiness, we need to look within ourselves to find it.

SEPTEMBER 25

Sniffing Out Cancer

A dog's nose rules its brain and interprets its world. Whereas a human has about 5 million scent receptors, a dachshund has 125 million and a bloodhound 300 million. We train dogs to sniff out narcotics, explosives, human remains, blood, wildlife feces, termites, bed bugs, and truffles. Malignant tumor cells produce volatile organic compounds, so could dogs be trained to sniff out cancer? In 1989, this question was posed in the medical journal *The Lancet* after a woman reported that her dog, a cross between a border collie and a Doberman, showed an inordinate interest in a skin sore on her body, which subsequently proved to be melanoma.

On September 25, 2004, an article in the *British Medical Journal* reported that dogs can be trained to distinguish between patients with bladder cancer versus those without bladder cancer on the basis of urine odor more successfully (41 percent) than would be expected by chance alone (14 percent). Since that study, dogs have been found to be capable of detecting lung and breast cancers from breath samples, ovarian cancer from tissue samples, colorectal cancer from breath and watery stool samples, and prostate cancer from urine. Because dogs' noses are so sensitive, they can sniff out cancers at very early stages. Trained dogs will never replace physicians, but they have proven themselves to be capable assistants.

Training and certifying dogs to detect cancer takes about 6–8 months. The dogs love their training, thanks to petting and play, treats, and other positive reinforcement. Little do they realize how they are transforming the field of cancer detection.

SEPTEMBER 26

Octopus's Hideaway

> We would be warm below the storm
> In our little hideaway beneath the waves
> Resting our head on the sea bed
> In an octopus's garden near a cave
>
> We would sing and dance around
> Because we know we can't be found
> I'd like to be under the sea
> In an octopus's garden in the shade.
> —THE BEATLES, "Octopus's Garden"

Thought to be the most intelligent of all invertebrates (and at least as smart as the average house cat), octopuses have long been admired. Ancient Greek and Roman artists embellished pottery, coins, and shields with the animals' graceful figures. In Ainu and Shinto lore from Japan, the huge octopus Akkorokamui can autotomize and regenerate its limbs, and thus is believed to possess healing powers. Believers offer fish, crabs, and mollusks at shrines that honor Akkorokamui in hopes of receiving cures for broken bones, mental purification, and spiritual release.

"Octopus's Garden," written and sung by Ringo Starr, was released on September 26, 1969. According to Ringo, "He [a ship cap-

tain] told me all about octopuses—how they go 'round the sea bed and pick up stones and shiny objects and build gardens. I thought, 'How fabulous!' because at the time I just wanted to be under the sea, too." In reality, octopuses take prey items like crabs and mollusks back to their dens where they eat them and leave the shells and other remnants strewn around their homes. These objects may help to camouflage and protect their hideaways, but in Ringo's imagination the decorative "gardens" are magical.

SEPTEMBER 27

Cry of Warning

> Like the constant dripping of water that in turn wears away
> the hardest stone, this birth-to-death contact with dangerous
> chemicals may in the end prove disastrous.
> —RACHEL CARSON, *Silent Spring*

Rachel Carson—ecologist, marine biologist, environmentalist, and nature writer extraordinaire—wrote several inspiring books about the sea before publishing *Silent Spring* on September 27, 1962. In it,

Carson warned about the long-term effects of misusing pesticides, and she called for a change in the way we view the natural world.

The final words in *Silent Spring* sum up Carson's message: "The 'control of nature' is a phrase conceived in arrogance, born of the Neanderthal age of biology and philosophy, when it was supposed that nature exists for the convenience of man. The concepts and practices of applied entomology [i.e., the use of pesticides] for the most part date from that Stone Age of science. It is our alarming misfortune that so primitive a science has armed itself with the most modern and terrible weapons, and that in turning them against the insects it has also turned them against the earth."

In all of her writing, Carson emphasized that humans are but one aspect of nature, yet one with power to alter the nature of the world. She encouraged us to appreciate our surroundings: "One way to open your eyes is to ask yourself, what if I had never seen this before? What if I knew I would never see it again?" If we value nature, perhaps we will no longer feel compelled to control her.

SEPTEMBER 28

Water Flowing Still Today

Mars, the "Red Planet," was named after the Roman god of war. Scientists have long speculated that Mars might have hosted life at one time, or even that life exists there today, because of potentially habitable conditions, including liquid water.

NASA's Mars Reconnaissance Orbiter (MRO) has been circling Mars since 2006 with scientific instruments. Its cameras have captured spectacular photos. On September 28, 2015, NASA reported that planetary scientists studying photos taken by MRO have detected 100-meter-long streaks on Mars' slopes, evidence of hydrated minerals and water flowing downhill. The photos are not of frozen water or signs of ancient water, but of water flowing still today. The streaks appear to ebb and flow over time, running down steep slopes during warm seasons and fading during cooler seasons. The news thrilled scientists who have held to the mantra "Follow the Water" in their search for extraterrestrial life. Although salty, water flowing on the surface of Mars renews hope that at least in some past time life existed on the planet.

NASA's MRO has identified a possible landing site for the Mars 2020 mission, planned for launch in July or August of that year. The four goals for robotic exploration during the mission are (1) to determine whether life ever existed on Mars, (2) to characterize the climate, (3) to characterize the geology, and (4) to prepare for human exploration. Stay tuned for the next installment in the search for extraterrestrial life.

SEPTEMBER 29

The Honking of Geese

Today is Goose Day in Mifflin and Juniata Counties, Pennsylvania. The event has a circuitous history. In 480 CE, Pope Felix III created Michaelmas Day to be celebrated on September 29 to honor the Archangel Michael. By the fifteenth century, Michaelmas Day had become one of the "quarter days" in the British Isles—one of the four days each year (spaced three months apart, on religious festival days) when accounts were settled. Eating a goose on that day was believed to protect against financial need, and the day became known as "Goose Day." In the 1780s, English settlers brought the belief—and the holiday—to Pennsylvania.

Geese are appreciated for more than their succulent flesh and ability to stave off bankruptcy. Domesticated grey geese make great watch animals, thanks to their territorial and aggressive behavior toward intruders. The goose was sacred to Juno, patron goddess of Rome. According to legend, the Capitoline Hill in Rome was saved from the invading Gauls in 390 BCE by the honking of Juno's geese. The Gauls had captured all of Rome except for Capitoline Hill. When

the Gauls approached, the geese quartered near the Temple of Juno cackled, hissed, and beat the Gauls with their wings. The commotion alerted the sleeping commander and his men, who drove away the Gauls. In honor of the geese, Romans long celebrated the event by an annual procession with a golden goose.

SEPTEMBER 30

Kissing the Sea

> I love oysters. It's like kissing the sea on the lips.
> —LÉON-PAUL FARGUE, French poet (1876–1947)

In 1954, hotelier Brian Collins obtained funding from Arthur Guinness to hold an Oyster Festival. His aim was to extend the tourist season of Galway, Ireland, into September. The event has run every year since, making it the world's longest running oyster festival. The Galway International Oyster & Seafood Festival has welcomed over 500,000 visitors who have consumed more than 3 million oysters, washed down with champagne, wine, and Guinness. The festival now spans four days on the last weekend of September.

Begin with the Guinness and Oyster reception. Listen to foodie talks, learn from cooking demonstrations, and tap your feet to tunes played by some of the best Irish musicians. Sample from the Seafood Trail of restaurants serving up gourmet dishes. Participate in the World Oyster Opening Championship. How long will it take you to shuck 30 oysters? (The current Guinness World Record holder shucked 233 oysters in three minutes.) If you really love oysters and have near-infinite capacity, sign up for the Tribal Oyster Feast Off (oyster-eating competition). Join the Mardi Gras Party with Champagne and stroll through the streets of Galway. Dance at the grand finale, the Gala Ball. If you need an added incentive, 30 Galway pubs offer free oysters with a pint of Guinness.

OCTOBER 1

Respect for Nonhuman Intelligence

> When our ancestors moved from hunting to farming, they lost respect for animals and began to look at themselves as the rulers of nature. In order to justify how they treated other species, they had to play down their intelligence and deny them a soul.
> —FRANS DE WAAL, *New York Times*, 2016

Today is Intelligence Day, in recognition of the October 1, 1946, founding of the Mensa Society, the world's largest and oldest high-IQ society. But let's not be anthropocentric. Let's also celebrate nonhuman intelligence today.

The more we study animal cognition (mental capacity), the "smarter" other animals appear to us. Adult male orangutans plan: they verbally announce which direction they expect to travel the next morning. Chimpanzees show empathy: they bring food to an injured companion and slow their walking pace so that the injured can keep up. Great apes, orcas, bottlenose dolphins, elephants, and Eurasian magpies recognize themselves in the mirror, suggesting

self-awareness. Clark's nutcrackers recall the location of thousands of seeds they buried a year earlier. Diverse animals, from cephalopods to primates, use tools.

Frans de Waal, an ethologist who studies animal cognition, asks: "How could our species arrive at planning, empathy, consciousness and so on, if we are part of a natural world devoid of any and all steppingstones to such capacities?" de Waal suggests that instead of insisting on our intellectual superiority, we should take pride in our connections with other animals.

OCTOBER 2

Mascots

"Brusha, brusha, brusha. Get the new Ipana! It's dandy for your teeth!" That catchy jingle will be a blast from the past for some readers. I have vivid memories of Bucky Beaver interrupting "Lassie Come Home" and "Sky King" during the 1950s to demand that Mr. Decay Germ stay away. You'll no longer find Ipana toothpaste in US grocery stores, but on October 2, 2009, the Canadian medical supply company Maxill purchased Ipana. The beaver is back.

Many companies use animal mascots. Geico Gecko declares, "This is my final plea: I am a gecko, not to be confused with GEICO, which could save you hundreds in car insurance. So STOP CALLING ME!" The Budweiser Frogs Bud, Weis, and Er croaked their names from the nighttime swamp, first randomly and then in sequence: "Bud-Weis-Er." Tony the Tiger proclaims that Kellogg's Frosted Flakes are "Grrrrrrrreat!" The pink Energizer Bunny claims that the Energizer battery "keeps going, and going, and going." Between 1997 and 2000 the Taco Bell Chihuahua woofed, "*¡Yo quiero Taco Bell!* (I want Taco Bell!)."

There's a reason animals are frequently used in marketing ads. Studies show that profits increase when animals are featured. We make an emotional connection to the product because the animals are memorable—cute, funny, or representing an honorable human attribute, such as loyalty, strength, or industriousness. When we see the product on the shelf, we think of the animal, smile—and, unhesitatingly, toss it into our cart.

OCTOBER 3

Blessing of the Animals

St. Francis of Assisi, Italian Roman Catholic friar and preacher, died on October 3, 1226. Many stories of Francis' life focus on his kindness toward animals. He preached to the birds and tamed a man-eating wolf. In remembrance of Francis' love of all animals, in early October, around the Feast Day of St. Francis on October 4, people worldwide take their pets—gerbils, rabbits, dogs, cats, iguanas, snakes, turtles, parrots, and more—to places of worship for the annual Blessing of the Animals. In Roman Catholic tradition, a prayer is offered and the pet is sprinkled with holy water. Many other religions, including Judaism and Unitarianism, have adopted this tradition of blessing pets, although St. Francis may not be a focus.

Most of us relate to the bond between humans and pets and the companionship they provide, and we applaud the therapeutic value of animals. But consider also the educational value of hamsters, guinea pigs, chicks, geckos, snakes, tadpoles, goldfish, tarantulas, and hermit crabs that visit the classroom. A 2014–2015 survey sponsored by the American Humane Association and Pet Care Trust of nearly 1200 teachers in North America revealed that teachers believe that class pets teach children empathy, respect, and responsibility for other living beings. Children learn social and leadership skills through caring for class pets. Pets teach children to give of themselves, an essential part of being human.

OCTOBER 4

Trading Debt for Conservation

Developing countries often convert their tropical forests into lumber for quick cash to pay off foreign debt. On October 4, 1984, conservation biologist Thomas Lovejoy wrote a *New York Times* editorial suggesting that reducing debt and promoting conservation could be done simultaneously through "debt-for-nature swaps."

The idea behind debt-for-nature swaps is that part of a developing nation's foreign debt is forgiven in exchange for investment in conservation. A conservation organization works with the debtor nation to develop a project, such as land protection. The conservation organization then purchases part of the loan at a discounted price from the creditor. Everyone gains. The debtor country reduces its external debt. The creditor gains and is willing to sell the loan at the discounted price because it assumes it will never see the debt fully repaid. The environment gains because the debtor country pledges to provide funds at an agreed-upon repayment schedule, to support the conservation project.

In 1987, Bolivia became the first country to participate in a debt-for-nature swap. Conservation International bought $650,000 of debt for $100,000. In exchange, Bolivia allocated funds for managing and protecting the Beni Biosphere Reserve and three large surrounding buffer zones. Debt-for-nature swaps have been carried out in other countries, including Mexico, Costa Rica, Ecuador, Peru, Brazil, Madagascar, Ghana, Zambia, and the Philippines. They offer hope to sustain long-term conservation efforts.

OCTOBER 5

Dogs 4 Diabetes

In October 2004, Mark Ruefenacht founded Dogs for Diabetes (D4D), a nonprofit corporation. Ruefenacht has type I diabetes. In 1999 when Ruefenacht had a severe hypoglycemic episode (an event during which blood glucose reaches a dangerously low level), his guide dog puppy-in-training aggressively aroused him from an incoherent state. Ruefenacht was able to treat himself and avoid becoming incapacitated.

Ruefenacht had professional experience in forensic science and knew of dogs' abilities to sniff out drugs, alcohol, and bombs. His own experience with his dog led him to speculate that perhaps dogs could be trained to smell and detect hypoglycemic episodes and alert people in danger. Ruefenacht worked for five years with other dog trainers in various disciplines, ranging from search and rescue to cancer detection. It turns out that diabetics give off a scent that signals hypoglycemia, and dogs can be trained to detect this scent and alert their handlers.

Since 2004, over 120 dogs—Labrador retrievers, golden retrievers, and other breeds—have been scent-trained by D4D. Once these dogs are paired up with diabetics, they sleep alongside the beds of their owners and accompany them during the day. Parents of diabetic children, and sons and daughters of elderly diabetic parents, can rest a littler easier knowing that scent-trained dogs are watching over their loved ones.

OCTOBER 6

Mad Hatter Day

October 6 is Mad Hatter Day! In 1986, a group of computer technicians in Boulder, Colorado, suggested that people might do less damage being silly for a day than by doing their jobs. They called it Mad Hatter Day, in celebration of Lewis Carroll's insanely silly character, the Hatter from *Alice's Adventures in Wonderland*. They chose October 6 because the book's illustration has the Hatter wearing a top hat with the words "in this style 10/6." Carroll actually meant for the numbers to refer to the hat's cost, but no matter. In Carroll's book, the Hatter attends a never-ending tea party with the perpetually late March Hare and singing Dormouse.

Instead, imagine a tea party with some of nature's real-life quirky animals. Observe a galaxy of starfish as they pry open clams with their arms, push their stomachs out from their mouths and onto the

exposed prey, digest, and then reclaim their stomachs. Leafcutter ants clip pieces of potted plants to take to their underground gardens where they will cultivate fungus. A pair of burying beetles eyes a dead mouse as potential food for their young. An army of soldier termites marches into the room. Frightened, the boxer crab stands ready to punch with sea anemones held in his claws. The three-lined potato beetle larva rolls in her toxic poop for protection, and the sea cucumber contracts his muscles and ejects his intestines. Real-life nature is even quirkier than Lewis Carroll's fanciful animals.

OCTOBER 7

Nevermore

> Quoth the Raven, "Nevermore."
> —EDGAR ALLAN POE, "The Raven"

Edgar Allan Poe's poem "The Raven" tells of a bird's mysterious visit to the distraught narrator lamenting the death of his lover, Lenore. The raven's response to each of the narrator's questions is "Nevermore," while the narrator slowly falls into madness. Poe died on October 7, 1849.

Poe followed a long line of writers who depicted the raven as a premonition of misfortune. In *Eclogues*, the Roman poet Virgil (70–19 BCE) repeated folklore that a raven observed from one's left side is an evil omen. In *Canterbury Tales*, Chaucer (ca. 1343–1400 CE) likewise wrote that a raven's presence is ominous. Almost everywhere the raven occurs, people have attributed it with supernatural powers. Its black plumage, raucous call, and habit of feeding on carrion reinforce it as an ominous symbol.

But ravens are also seen as beneficial. In ancient Greece, the raven was a symbol of wisdom and good luck. Babylonian mariners released ravens from their ships and watched the birds' responses. If the birds flew purposefully in a given direction, the mariners knew where to find land. In the Yukon, First Nations peoples performed a raven dance to please the ancestral spirits so they would send game. For the Haida and Tlingit of the US Pacific Northwest, the raven was a culture hero. In shamanic philosophy, there is "good" and "bad" in every being. And so it is with ravens.

OCTOBER 8

Leaping over Waterfalls

Imagine an animal whose express goal in life is a single act of reproduction. To accomplish that, the animal must leave the ocean and return to its birthplace to spawn in fresh water. It fights its way upstream, leaping over waterfalls. Getting knocked back downstream, it tries again. Within days or weeks after spawning, the animal dies, expended of energy, a symbol of endurance, self-sacrifice, and loyalty to home. We call this animal salmon, from the Latin word *salire*, meaning "to leap." In 2015, Chicken of the Sea announced the creation of National Salmon Day, to be celebrated annually on October 8. Statistics for 2017 indicate that salmon is the second-most-consumed seafood (behind shrimp) in the US, but the red fish has provided people critical protein since long before it was touted as a high-omega-3 food.

Some indigenous peoples perform first-salmon ceremonies to ensure the fish's perennial return. For example when a Kwakiutl fisherman from Vancouver Island, British Columbia, lands his first sockeye salmon of the season, he removes the spear and prays to the fish. He asks for protection for his family and requests that the fish send its own relatives so that the fisherman might spear them. The fisherman's wife joins her husband on the beach where she also prays to the fish before roasting it for a guest of honor. If the roasted eyes are not eaten that night, sockeye salmon will disappear forever. After the fish has been eaten, the wife wraps the skin and bones in a food mat and tosses the bundle into the ocean so that life will be rekindled.

OCTOBER 9

Swan Song

The French Romantic composer Camille Saint-Saëns was born on October 9, 1835. One of his best-loved pieces is the fanciful "Carnival of the Animals." Each movement represents a different animal or group of animals, portrayed by musical instruments that reflect the animals' personalities and characteristics.

We begin with strings and two pianos—a bold tremolo announcing lions waking, stretching, and roaring. Next, violins imitate clucking chickens, and a clarinet echoes a crowing rooster. Two pianos playing scales up and down the keyboard represent galloping jackasses. Strings and piano create a slow rendition of the "Cancan" to portray the turtle. Low, plodding sounds from the double bass and piano mirror a clumsy, dancing elephant. And so it goes, with kangaroos, fish, donkey, birds, and even skeletons joining the carnival.

Second-to-last is "The Swan," the most famous of the movements and the only one Saint-Saëns allowed to be performed or published before his death. A solo cello depicts a swan gliding over the water, and two pianos echo the created ripples. Ancient Greeks and Romans (incorrectly) believed that the mute swan is silent until its final moments of life, when it sings the most beautiful of all songs—its "swan song." This lusciously romantic movement reflects the ultimate "swan song."

We end the piece with the full ensemble of instruments—all the animals in the carnival. Just before the final chords, Saint-Saëns gives us six "Hee-Haws." The jackass has the last laugh.

OCTOBER 10

Herb of the Sun

Marigolds, the birth flower for October, represent warmth and fierceness, elegance and devotion. For centuries, herbal healers have valued marigold flower heads as both internal and external medicaments, including treatments for fevers, jaundice, warts, and bee stings. Botanist, herbalist, and physician Nicholas Culpepper (1616–1654) referred to marigolds as an "herb of the sun" and advocated

a plaster of dried marigold flowers, hogs' grease, turpentine, and resin applied over the breast to strengthen the heart.

Joyous celebrations deserve vibrant flowers, and so cultures worldwide feature bright gold and orange marigolds in their festivities. Late October/early November sees the 5-day Hindu celebration of Diwali, the Festival of Lights. Traditional customs include illuminating one's home with candles and oil lamps and decorating with marigolds—garlands of marigolds hung from railings and balconies, runners of marigolds on the dining tables, and carpets of marigold petals by the front door. From October 31 to November 2, Mexicans celebrate Dia de los Muertos (Day of the Dead), a gala celebration of life and a time of calling back the dead for remembrance. Loved ones set up colorful altars featuring the deceased's favorite foods, personal objects, and marigolds, considered to be flowers of the dead. In some villages, people leave trails of gold and orange marigolds from their front doors to loved ones' graves. The strong fragrance of the flowers is said to guide the spirits of the dead back home to their altars.

OCTOBER 11

America's First Great Conservation Success

An estimated 30–60 million American bison once dominated what would become the United States of America. These giant herbivores helped sustain plains and prairie ecosystems, and they shaped the lifestyle of many tribes of Native Americans. As settlers expanded west, they hunted bison for sport. They also slaughtered bison for their meat to feed railroad crews, for their bones to make fertilizers and fine bone china, and for their hides to provide robes. The vast herds of bison had dwindled to 1000 individuals in the US by the late 1880s.

In 1905 William Hornaday, Theodore Roosevelt, and others founded the American Bison Society to restore bison populations.

On October 11, 1907, 15 bison captive-bred at the Bronx Zoo were shipped by railroad to the Wichita Mountains Wildlife Refuge in southwestern Oklahoma. By this time, bison had been extinct on the southern Great Plains for 30 years. The newcomers thrived and reproduced.

Although there are an estimated 340,000 bison in the country today, most are the result of cross-breeding with cattle and are semi-domesticated. Truly free-ranging herds occur only in Yellowstone National Park (about 3500 individuals) and in southern Utah in the Henry Mountains and Book Cliffs (about 500 individuals combined). Repatriation of American bison is recognized as America's first great conservation success, with the animal no longer listed as Endangered. The work is not finished, however. More than 60 Native American tribes are working to restore free-ranging bison to their original prairie habitats in the West and Midwest.

"The Skunk"

Every year in mid-October the Nobel Prize in Literature is announced. In 1995, that honor went to Seamus Heaney, probably the best-known and loved Irish poet since William Butler Yeats. One of Heaney's celebrated poems is "The Skunk," published in the poetry collection *Field Work* (1979), a love poem in which a skunk reminds Heaney of his wife. On first hearing this metaphor, you might assume the comparison would be unflattering. Instead, it reflects devotion and tenderness.

Heaney spent time as a guest lecturer at the University of California, Berkeley, in 1970–1971. During that time, he profoundly missed his wife back in Ireland. "The Skunk" was inspired by a skunk Heaney fed each evening when it visited his garden. In the second stanza of the poem, reflecting his emotions while waiting for the skunk to arrive, he wrote: "I began to be tense as a voyeur." In the

last stanza, Heaney compared his waiting for the skunk to waiting in bed for his wife to change into her nightclothes. Following are the first and last of the six stanzas of "The Skunk."

> Up, black, striped and demasked like the chasuble
> At a funeral mass, the skunk's tail
> Paraded the skunk. Night after night
> I expected her like a visitor.

> It all came back to me last night, stirred
> By the sootfall of your things at bedtime,
> Your head-down, tail-up hunt in a bottom drawer
> For the black plunge-line nightdress.

OCTOBER 13

Meditating in Nature

The October 13, 1975, cover of *Time* magazine featured Maharishi Mahesh Yogi, the man who developed the technique of Transcendental Meditation (TM). The inside article was entitled "The TM Craze: 40 Minutes to Bliss." Medical experts have confirmed many potential benefits of meditation, including TM. Meditation lowers blood pressure and increases energy level. Meditation deceases tension, stress, and anxiety, and increases creativity and emotional stability. Maharishi Mahesh Yogi became the guru of transcendental meditators worldwide, many of whom advocate meditating in nature.

There's a reason that many meditation centers and monasteries are located in isolated mountain and forest surroundings. Untamed nature awakens our senses, heightens our sensitivity, opens our minds, and encourages spiritual growth. Nature meditation doesn't have to be in a spectacular, tranquil setting, however, and you don't have to join an organized retreat. Try meditating in your backyard, in a park, or a vacant lot. Find a comfortable spot—boulder, log, grass, or lawn chair. Take several deep breaths and then focus on your breathing. Be aware of your surroundings as though you had never experienced them before. Delight in the beauty of nature. Feel the breeze on your face and the sun on your back. Tune in to the bird songs, the silence. Smell the air, the earth. Melt into the landscape.

OCTOBER 14

Nature's Music

October is National Protect Your Hearing Month. Close your eyes and listen to nature—from breezes blowing through a quaking aspen grove, raindrops plunking on a lake surface, and the whoosh of a great blue heron lifting off the ground, to thunder booming and lightning crackling, ocean waves crashing against the shore, and hot gasses roaring as they escape a vent during a volcanic eruption. All are part of nature's music.

Hearing likely evolved for detection of predators. Many animals have better hearing than humans' average range of about 20 Hz (hertz) to 20 kHz (kilohertz). Dogs can hear from about 67 Hz to 45 kHz; cats 55 Hz to 79 kHz; and bats 1 to 212 kHz. Bats use ultrasonic calls to locate moths. Although many moths can hear some high ultrasonic bat calls, until recently none was known to detect the highest frequencies used by bats. The record was held by the North American gypsy moth at 150 kHz. Then in 2013, Hannah Moir and her colleagues reported that the greater wax moth (*Gallenia mellonella*) can detect ultrasonic frequencies approaching 300 kHz, giving it the highest frequency sensitivity of any animal. It's an evolutionary arms race. As bats make their calls less detectable, moths increase their ability to detect high frequency calls. At the moment, moths—at least greater wax moths—may be winning.

Spider Silk without Spiders

E. B. White's story *Charlotte's Web*, published October 15, 1952, tells of friendship between Wilbur the pig and a barn spider named Charlotte. When Charlotte hears that Wilbur is to be slaughtered, she spins messages in her web praising Wilbur to attract support for her friend.

Arachnologists and author E. B. White are not alone in admiring spider silk—a substance 100 times stronger than human ligaments, 10 times stronger than tendons, pound for pound stronger than steel, and more elastic than nylon. What if we could use spider silk to manufacture our own materials? Because spiders tend to be territorial and cannibalistic, we cannot raise them communally to collect large quantities of silk.

Instead, scientists and biotechnologists are creating synthetic spider silk. At Utah State University, Randy Lewis and his colleagues use transgenic (genetically altered) goats, *E. coli* bacteria, transgenic alfalfa, and transgenic silk worms to make synthetic spider

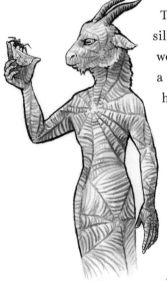

der silk proteins that are then spun into fibers. The researchers are working to mass-produce silk to make a stronger and lighter fabric that would replace nylon for Army clothing. Spiber, a Japanese advanced biomaterials company, has paired with the outdoor recreation company North Face and made a "Moon Parka" from faux spider silk—tough and lightweight, selling (so far only in Japan) for $1000. Someday, ligaments made from synthetic spider silk may give people with injured knees or shoulders renewed mobility, and bulletproof vests constructed from the material may protect law enforcement officers. The possibilities are limited only by our imaginations.

Let Your Moose Loose!

Long, spindly legs support an ungainly brown body. The front end is ornamented with weighty headgear, an oversized, fleshy, drooping schnoz, and a flap of skin dangling from the throat. The back end is punctuated by a short, stubby tail, which doesn't seem particularly functional. It's the moose, the world's largest member of the deer family and a cultural symbol of wilderness, imbued with mystique.

Minnesotans love their moose. The third weekend in October is the Moose Madness Family Festival in Grand Marais, Minnesota, timed to coincide with moose mating (rutting) season. Moose go a little crazy this time of year. Bulls grunt and bellow to attract females, and they fight other males for access to females. Sometimes these battles turn into epic fights lasting several hours, with bulls locking antlers and ramming each other. A large bull can weigh more than 1200 pounds (545 kg) and sport antlers that span over five feet (1.5 m). Two big guys fighting, each pumped up with testosterone, is indeed "moose madness."

There's a lot to do during the three-day festival. In addition to moose-related lectures and discussions with wildlife experts, you can participate in the Moose Tracks Race (decked out in your best moose-themed apparel), guess the number of moose droppings in a jar, enter the moose cartoon contest, or try your hand at writing moose-themed haiku. Let the games begin, or as they say in Grand Marais, "Let your moose loose!"

OCTOBER 17

Tribe of Hairy Women

Huge hairy animals, seemingly part human, part ape, created terror in the tropical and subtropical forests of central Africa. Locals viewed the creatures as cannibals endowed with magical powers. Western scientists doubted that such an animal existed, even in benign form—until 1847. That year, Reverend Thomas Savage, a trained physician and American missionary in West Africa, was shown some bones and an unusual skull of an animal described by locals as a "large, monkey-like animal" collected in Gabon. He recognized it as a new species of primate and sent the bones to Jeffries Wyman, an anatomist at Harvard University. Wyman named the new primate *Troglodytes gorilla* Savage, giving Savage credit as its describer. [The species is now *Gorilla gorilla* (Savage)].

The name "gorilla" comes from an account by a Carthaginian navigator and explorer named Hanno, who sailed along the west coast of Africa nearly 2500 years ago. He reported seeing a group of hairy, human-like creatures, most of whom were females. His interpreters said the locals referred to these animals as *gorillai*, a word (language unidentified) that loosely meant "hairy person." We get the word gorilla from Hanno's Greek translation of *gorillai*, meaning "tribe of hairy women."

On October 17, 1902, German army officer and explorer Captain Robert von Beringe discovered an even larger gorilla on the ridges of the Virunga Mountain range—the mountain gorilla. Although mountain gorillas (*Gorilla beringei beringei*) are intimidating because of their immense size and strength, coupled with chest-beating displays by males, George Schaller and Dian Fossey showed us the gentle nature of these magnificent primates.

OCTOBER 18

Due More Respect

Before children can even ride a tricycle they've already been indoctrinated about mean, scary wolves from "The Three Little Pigs" and "Little Red Riding Hood." To dispel public misconceptions about these long misunderstood and persecuted animals, in 1996

Defenders of Wildlife established the third week of October as National Wolf Awareness Week in the United States.

At one time, gray wolves roamed over most of the lower 48 states. By the early 1900s, there were few wolves left; they had been victims of unrestricted hunting and government extermination programs. Within the past several decades, thanks to protection and reintroduction, gray wolves have made a comeback in some areas (e.g., the Great Lakes region and Yellowstone), but they are still largely met with intolerance and fear.

Cave paintings suggest that early humans respected wolves. We share a lot in common with wolves' social behavior—living in family groups, maintaining pair bonds, and hunting cooperatively. Long ago, wolves may have enjoyed the warmth of our fires. In turn, we may have benefitted from early warning of danger through wolves' barking and howling. The relationship went horribly wrong, however, once humans developed agriculture and began domesticating animals. At that point wolves preyed on livestock. Humans and wolves competed and became adversaries. We have the upper hand now. We determine where wolves can live and how many can live in those places. As key components of ecosystems, gray wolves deserve more respect than they get. We need, once again, to learn how to live alongside wolves.

Rainforest Day

Do you have a favorite ecosystem? Mine is tropical rainforest. In an article entitled "Evolution in the Tropics," evolutionary biologist Theodosius Dobzhansky wrote: "Tropical life seems to have flung all restraints to the winds. It is exuberant, luxurious, flashy, often even gaudy, full of daring and abandon, but first and foremost enormously tense and powerful." In my travel/adventure memoir *In Search of the Golden Frog*, I reflected on Dobzhansky's words, recalling my own impressions of the rainforest: "*Exuberant*: two-meter-wide giant lily pads, strong enough for children to lie on. *Luxurious*: green on all sides, from mosses to canopy trees. *Flashy*: iridescent purple, green, and blue hummingbird feathers shimmering in the sunshine. *Gaudy*: turquoise, black, and orange poison dart frogs. *Daring and abandon*: raucous squawking of parrots flying over the canopy. *Tense and powerful*: emerald tree boas squeezing life from birds before swallowing them whole." Now, 50 years after my first trip to the tropics in 1968, the rainforest still holds an irresistible allure for me.

October 19 is Rainforest Day, an occasion used to raise awareness of rainforest destruction. Whereas rainforests once covered about 14 percent of Earth's land surface, now they cover 6 percent. Forests are being cut down for cattle ranching, agriculture, and logging and mining operations. Experts warn that the world's rainforests may be gone by 2065. Such a loss will have major repercussions. Rainforests generate 28 percent of the world's oxygen turnover. More than half of all species of plants and animals live in the rainforest. How could we live without rainforests?

International Sloth Day

In the late 1700s, West Virginia saltpeter miners working in a cave dug up long, slender leg bones, and foot bones with vicious-looking claws. They sent the fossils to Vice President Thomas Jefferson, who was fascinated by science and particularly intrigued with paleontology. In 1797, Jefferson described the fossils in a paper he read at a meeting of the American Philosophical Society. Jefferson suggested the name *Megalonyx*, meaning "great claw," and thought the animal to be some sort of lion or other wild cat. Although scientists were becoming convinced that extinction played a major role in life on Earth, Jefferson adhered to "completeness of nature," the belief that the inherent balance of nature prevented natural extinction. He thought *Megalonyx* might still live, and he instructed Lewis and Clark to keep an eye out for it during their 1804–1806 expedition.

The animal, later named *Megalonyx jeffersonii*, turned out to be a giant ground sloth, a Pleistocene vestige from the most recent Ice Age in eastern North America.

October 20 is International Sloth Day, created in 2010 by the Colombian conservation and wildlife organization AIUNAU. Sloth, the attribute, is one of the seven deadly sins of early Christian origin (along with pride, greed, lust, envy, gluttony, and anger). Sloths, the animals, were named after the deadly sin because of their slow and seemingly lazy nature. Today we have six species of sloths, all living in Central and South America, and all a fraction the size of the 9.8-foot (3-m) "Jefferson's giant claw."

OCTOBER 21

Save the Geckos!

October 21 is National Reptile Awareness Day, started by US reptile enthusiasts who wanted to change public perception of these animals. On this day, individuals and organizations focus on outreach to promote public education, conservation, and appreciation for the world's more than 10,000 species of reptiles.

Public outreach can make a difference. One example is a project called "Save the Geckos!" In southern Portugal, popular folk belief maintains that two species of geckos, lured by insects attracted to lights around people's homes, are venomous and carry a nasty skin disease. In 2010, biologist Luis Ceríaco and his colleagues surveyed 865 people and discovered that 22 percent kill geckos they encounter. Another 8 percent ask someone else to dispatch the lizard. Ceríaco and his colleagues wrote a children's book about a little boy and his pet gecko. The book dispels the erroneous folklore about venom and skin disease, and tells of a plague that happens when geckos stop eating mosquitoes. The biologists read the story to 114 kindergarten and primary school children. A week before hearing the story, the children were asked to draw a gecko and describe it with five adjectives. After hearing the story, the children were asked to draw another gecko and describe the animal with five adjectives. Children's attitudes toward geckos improved dramatically. Before hearing the story, most students used adjectives like dangerous and ugly; afterward, they described the geckos as useful and friendly. Hopefully those children, unlike their parents, will appreciate and protect the local geckos.

OCTOBER 22

Wombat Day

If you've ever felt inspired to make a wombat-shaped chocolate cake, cookies, brownies, or fudge, today is the day. Aussies will be baking and consuming them today, for October 22 is Wombat Day, celebrated since 2005.

Koalas, wombats' closest relatives, might rank higher on the cuteness and cuddly scales, but wombats—pudgy, muscular, and stubby-legged—have their own allure. They are unconventional. For starters, wombats have very slow metabolism and very long intestines. They take 8–14 days to digest a meal of grasses and roots. Most of the moisture is absorbed in the digestive tract, resulting in compact, cubic turds. Wombats deposit these poop cubes around their burrows to mark their territories. They spend a lot of time

inside their burrows—about 75 percent of their lives. Wombats are powerful excavators, able to out-dig a person with a shovel. Thanks to a backward-opening pouch, a digging mama wombat doesn't cover her pouch-riding baby with dirt. Wombats function like miniature bulldozers, using their flat foreheads as battering rams as they charge through brush, fences, and screen doors.

Celebrate today with a wombat party—eat chocolate, share wombat folktales, sing wombat songs, and, if you can find someone to teach it to you, dance the wombat dance. According to the website Wombania, the entire chocolate wombat cake must be eaten during the party, or your wombat wishes will not come true.

Creation

James Ussher (1581–1656), biblical scholar and Archbishop of the Church of Ireland, is best known for the "Ussher Chronology," 2000 pages of text proposing, based on a literal reading of the Old Testament, that the world was created at midday on October 23, 4004 BCE.

In the seventeenth century, 4000 BCE was a commonly accepted Bible-based estimate for the creation of the world. The world's potential duration was assumed to be 6000 years—4000 before the birth of Christ, and 2000 years afterward. But chronologists had already established that because King Herod, who had ordered the slayings of all young male children in and around Bethlehem, died in 4 BCE, Jesus could not have been born after that year. Thus, Ussher used 4004 BCE as the year the world was created rather than 4000. The Jewish year began in autumn, and October was the autumnal equinox using the old Julian calendar. God purportedly created in six days and then rested on the seventh; the Jewish Sabbath is Saturday, so God must have started on a Sunday. Ussher designated the 23rd of October because it was the first Sunday following the autumnal equinox in 4004 BCE. Ussher set the creation of light at high noon for no given reason.

In a 1991 essay in *Natural History* magazine, evolutionary biologist Stephen Jay Gould did not defend the substance of Ussher's chronology, but rather argued that the chronology was "an honorable effort for its time." Gould ended his essay with a plea for "judging people by their own criteria, not by later standards that they couldn't possibly know or assess."

OCTOBER 24

Bird Feathers Spark a Conservation Movement

> The Illinois Audubon society is trying to save our wild birds from extinction. We purpose [*sic*] doing this by pledging women not to wear birds or bird plumage of any kind except ostrich plumes on their hats. Ostrich plumes are exempted because their gathering does not involve the torture or killing of the birds.
>
> —*Chicago Daily Tribune*, October 24, 1897

During the nineteenth century, taxidermied bluebirds, Baltimore orioles, redheaded woodpeckers, and hummingbirds decorated women's hats, along with plumes of snowy egrets, white ibises, great blue herons, birds of paradise, pheasants, and ostriches. By the mid- to late 1800s, millions of birds were killed worldwide each year for women's millinery fashion. In its winter 1886–1887 issue, *Good Housekeeping* magazine reported: "At Cape Cod, 40,000 terns have been killed in one season by a single agent of the hat trade."

By the late 1890s, women conservationists who were concerned about bird declines held tea parties to convince their wealthy friends to boycott the feather millinery business. In 1896, outraged by the slaughter of birds to adorn women's hats, Boston socialites Harriet Hemenway and Minna Hall founded the Massachusetts Audubon

Society for the protection of birds. By 1898, Audubon societies had been established in 15 other states and the District of Columbia. Urged by the Massachusetts Audubon Society, the first federal-level conservation legislation was passed—the Lacey Act of 1900.

OCTOBER 25

Gastronomic Treasures

How do you like your truffles? Shaved over steaming pasta, steak, fish, or shrimp? Infused in oil, honey, or vodka? Baked into chocolate cake? Truffles are the fruiting bodies of certain fungi that grow underground, usually associated with tree roots. These fruiting bodies, offering pungent aroma and strong, earthy flavor, are one of the world's gastronomic treasures. White truffles from Italy will cost you $125/ounce—or more. Perigord black truffles from France hover around $35/ounce. Ounce for ounce, some truffles are one of the world's most expensive foods, up there with saffron and caviar.

The ancient Romans appreciated truffles and believed they grew because Jupiter, king of the gods, threw a lightning bolt onto his sacred oak tree. Jupiter was renowned for his sexual exploits, which explains why truffles have long been touted as aphrodisiacs.

People often enlist dogs and female pigs, because of their keen sense of smell, to hunt truffles. Dogs must be trained for the job, but female pigs have a natural instinct to dig up truffles because they are strongly attracted to a compound in truffles similar to androstenol, the sex pheromone found in boar saliva. An advantage to truffle hunting with a dog is that it can be trained to relinquish its find. It's much harder to keep a pig from devouring its unearthed treasure.

Foodies will delight in the international White Truffle Festival held annually in Alba, northern Italy, from early October to mid-November, a festival celebrated for nearly 90 years. Festivalgoers can join a truffle hunt, buy fresh truffles at the market, and graze from food stands selling truffle-infused delicacies.

OCTOBER 26

Royal Gift

George Washington recognized the value of mules (sterile hybrids between male donkeys and female horses) in agriculture. He wanted to breed them, but he needed high-quality donkey stock. American donkeys, descendants of animals brought by early explorers, were small. Andalusian donkeys from Spain—large, strong, and calm—were considered the best of the best, but Spain prohibited their exportation to maintain control. Four years before becoming President, Washington wrote directly to King Charles III of Spain asking to buy high-quality breeding stock. On October 26, 1785, a ship docked in Boston harbor with a gift from King Charles: one jack (male donkey) appropriately named Royal Gift, and two jennies (female donkeys). The following year, the Marquis de Lafayette of France sent Washington a Maltese jack and several jennies. These

mated with the Andalusian donkeys and produced a strong breeding stock to cross with horses.

Washington became America's first mule breeder, and Royal Gift is credited with beginning an American dynasty. Mules have the best traits of both parents: athletic ability, courage, and speed from their horse mothers, and strength, intelligence, sure-footedness, and endurance from their donkey fathers. Mules replaced horses on farms. They pulled pioneers' wagon trains and stagecoaches. They hauled supplies to gold mines and transported the gold to banks. By 1808, America had an estimated 855,000 mules. Under the Reagan administration, in 1985 a bill designating October 26 National Mule Day was signed into effect to mark the 200th anniversary of the gift that helped shape America.

OCTOBER 27

The Humble Potato

October 27 is National Potato Day in the United States. We have tantalizing ways to prepare and eat potatoes: garlic mashed, twice-baked, oven-roasted, au gratin, scalloped, potato chowder, German potato salad, hashbrowns, potato chips, and French fries. Americans, on average, consume 132 pounds (60 kg) of potatoes per person per year. In 2011, we collectively consumed an estimated 1.5 billion pounds of potato chips! Europeans are the world's heartiest potato-consumers, at 194 pounds (88 kg) per person each year.

Potatoes have a long and convoluted history. Genetic evidence reveals that the earliest potatoes were cultivated in the Andean highlands of Bolivia and Peru, at least by 3400 BCE. The Inca ate potatoes as nourishment and to prevent indigestion; they placed slices of raw potatoes on broken bones to promote healing; and they buried potatoes with their deceased. In the mid-1500s, Spanish conquistadors carried potatoes back to Europe from South America. In 1621, British colonists introduced potatoes to eastern North America. During the late 1890s, Alaskan Klondike Gold Rush miners so valued potatoes that they traded gold for potatoes ounce for ounce. Coming full circle, today more than 2800 varieties of potatoes are grown in Peru. Potatoes produce more weight of food per area of cultivated land than any other major planted crop, in itself a reason to celebrate the humble potato.

OCTOBER 28

Magnificent Flying Machines

Birders from around the world flock to Viña del Mar on the coast of central Chile during four days in late October for the annual Birds of Chile Festival, to birdwatch and add exotic species to their "life lists." Chile has a diverse avifauna consisting of about 500 species, including three species of flamingos, 10 hummingbirds, the lesser rhea, six tinamous, 11 albatrosses, the wattled jacana, five parrots, two storks, three tropicbirds, two frigatebirds, six boobies (including blue-footed and red-footed), and many ducks and other waterfowl, just to mention a few.

Perhaps most sought-after by birders is the Andean condor (*Vultur gryphus*), the national bird of Chile (and also of Bolivia, Colombia, and Ecuador). The Andean condor is the largest flying bird in the world by a combined weight and wingspan measurement—truly a magnificent flying machine. These birds can reach 33 pounds (15 kg), with a wingspan of 11 feet (3.3 m). Andean condors nest at elevations up to 16,000 feet (nearly 5000 m). For the Inca, Andean condors represented *hanan pacha*—the upper realm of the sky, sun, moon, and stars. Andean condors were, and still are, revered by Quechua and other Andean cultures as rulers of the Upper World. Today, an estimated 10,000 Andean condors fly free in the wild and continue to inspire us as a symbol of strength, power, and freedom.

OCTOBER 29

Picky Eaters

October 29 is National Cat Day in the US, an excuse for cat owners to lavish love and attention on their pets. Many owners resist the temptation to give their cats special treats, however, for cats generally are unwilling to try new foods. Most are picky eaters. There's a good reason for this. Cats have evolved to eat a nearly all-meat diet. In the wild, eating a new food could sicken a cat, so it's safer to stick with the familiar.

In 2016, Adrian Hewson-Hughes and his colleagues published research showing that domestic cats are driven to eat foods with a preferred fat-to-protein ratio; flavor is secondary. First the investigators offered cats wet food options of the same fat-to-protein ratio, with three formulated flavors: fish, rabbit, and orange. The cats preferred fish, then rabbit; as you would expect, orange came in a distant third. Next, cats were given their choice of nine options: three flavors (fish, rabbit, and orange), each offered in three fat-to-protein ratios (90, 60, and 30 percent fat options). Over the long term, the cats selected meals based on nutritional content, preferring the 30 percent fat options, regardless of flavor—even orange! How cats detect the ideal nutrient ratio remains a mystery. But then, as every cat owner knows, cats are mysterious animals.

OCTOBER 30

A Celebration of Animals

Diwali, "Festival of Lights," is a joyous five-day celebration by Hindus, Jains, Sikhs, and some Buddhists of the triumph of light over darkness, good over evil. Diwali occurs two days before, the day of, and two days after the new moon of the Hindu month of Kartika (mid-October to mid-November in the Gregorian calendar). The celebration includes the lighting of millions of lights, gift-giving, fireworks, story-telling, singing, dancing, and feasting.

In Nepal, where the Hindu celebration is called Tihar, celebration of animals is a major focus. The first day of Tihar is dedicated to crows and ravens, considered to be messengers of death because of their sad cawing. To avert death and grief for the coming year, sweets and other foods are placed on roofs to propitiate the

birds. On the second day, dogs are fed treats, decorated with garlands, and appreciated for the joy they bring. Cows, considered sacred, are honored on the third day, decorated with garlands and fed high-quality grass. Depending on their cultural background, on the fourth day Nepalese celebrate oxen (representing the earth), cow dung (representing the Holy Goverdhan Hill), or themselves, an occasion to purify their bodies. On the last day of Tihar, sisters honor their brothers. No one is left out of the day. Those without sisters or brothers celebrate with relatives. Friends, family, and the animals with which we share intense relationships are ultimately what life is all about.

OCTOBER 31

Truth of Imagination

> I am certain of nothing but the holiness of the heart's affections
> and the truth of imagination—what the imagination seizes
> as beauty must be truth—whether it existed before or not.
> —JOHN KEATS, from a letter written to Benjamin Bailey, 1817

John Keats, beloved English Romantic poet, was born on October 31, 1795. He died of tuberculosis at age 25. As his disease progressed, he became obsessed with the value and necessity of imagination to escape reality. Keats profoundly revered this unique and special quality of the human mind, and is often called "The Poet of Imagination."

Think about how we view nature through our imagination. Depending on your culture, you might see a man in the moon, or a rabbit, frog, buffalo, or dragon. We see human profiles and reclining figures, eagles' beaks, and razorbacks in rock formations. We gaze at the cloud-studded sky and see celestial flowers, swimming fishes and turtles, human faces, elephants with upturned trunks, and fluffy white rabbits. A saguaro cactus might resemble an octopus, a giant garden cultivator, or a starfish impaled on a stick. We see

faces in tree bark and in the blemishes on fruit and vegetable skins. Pareidolia refers to the phenomenon whereby the human mind perceives a familiar pattern or image where none actually exists. Many, perhaps most of us, experience pareidolia.

NOVEMBER 1

Honor the Bison

> It is now the season at which the buffaloe begin to coppelate and the bulls keep a tremendous roaring we could hear them for many miles and there are such numbers of them that there is one continual roar. . . . The missouri bottoms on both sides of the river were crowded with buffaloe I sincerely belief that there were not less than 10 thousand buffaloe within a circle of 2 miles arround that place.
> —MERIWETHER LEWIS, journal entry July 11, 1806 (north-central Montana)

The first Saturday in November is National Bison Day, an event celebrated annually since 2012. It is a day to honor the largest living land animal in all of the Americas. The American bison (often called buffalo) serves as the state mammal for Kansas, Oklahoma, and Wyoming. It has appeared on the official seal of the US Department of the Interior almost continuously since 1917. In May 2016, President Barack Obama signed the National Bison Legacy Act, a bill designating the American bison as the official national mammal of the United States. The designation honors the animals' cultural, ecological, historical, and economic contributions to America's heritage.

The bison is the iconic symbol of the free and open spirit of the American West. It shaped the ecology of the Great Plains and the lives of the Plains Indians. More than this, the bison reminds us of our ability to drive other species to near-extinction, and of our responsibility to conserve our country's wildlife and their remaining wild areas.

NOVEMBER 2

Released as a Butterfly

The ancient Greeks believed that when the soul (Psyche) left the human body at death, it was released as a butterfly. The Aztecs believed that their deceased relatives revisited Earth as butterflies to assure the living that all was well.

Various cultures still associate butterflies with departed human souls, a belief that may have been handed down from the ancient Greeks and Aztecs. Every early November, monarch butterflies migrate from Canada and the United States to the mountains of central Mexico. Many local people view the butterflies as souls of the deceased returning for a visit. The black, orange, and white monarchs add a splash of color, as well as symbolism, to the All Souls' Day festivals held on November 2. In Romania, people open a window or door after a person dies, so the soul can leave the body as a butterfly. In Andalusia, Spain, loved ones splash wine over a deceased's ashes—a toast to the butterfly that will escape when the soul leaves the body. Butterflies are commonly seen around the flowers left at gravestones. A butterfly in flight sometimes hovers for a few seconds, appearing hesitant—a soul reluctant to proceed to the next phase. Whether we truly believe that butterflies are human souls or we relate to the association as a powerful metaphor, butterflies symbolize rebirth.

NOVEMBER 3

Keep on Keepin' On

One day about 310 million years ago, a jellyfish the size of a smartphone found itself beached on an ancient ocean shore. As it floundered, it pumped sand into its body. Storm winds or a large wave deposited the sand-filled corpse in an oxygen-deprived lagoon where it sank into the mud. Over time, silt and clay hardened around the jellyfish, forming black shale. This blob of white sand encased in shale, complete with trailing tentacles, was discovered in an Indiana quarry in 1973. Four decades later, paleontologist Graham Young and a colleague reconstructed this story of the jellyfish's final moments and subsequent preservation.

Jellyfish are 95 percent water. Without any bones or other hard body parts, they leave little behind when they die. If they get buried quickly in fine sediment, however, they can leave imprints in the rock. This type of fossilization explains what happened to the oldest known fossil jellyfish, from the Cambrian Period in present-day Utah, about 505 million years ago.

November 3 is Jellyfish Day, recognized by the National Museum of Animals and Society in Los Angeles, California. Why honor jellyfish? Why not! They've earned our respect for having survived a half billion years. They were around long before the early ichthyosaurs and other huge reptiles terrorized ocean life 250 million years ago. Jellyfish saw these Jurassic giants go extinct, but they kept floating in the ocean currents and swimming by jet propulsion. "Jellies," the oldest multi-organ animals on our planet, are found in every ocean today.

NOVEMBER 4

Protection for Afterlife

King Tutankhamen became Pharaoh of Egypt at the tender age of nine or ten. He died at about 19 years old, his reign lasting from about 1332 to 1323 BCE. English archeologist and Egyptologist Howard Carter discovered Tutankhamen's tomb in the Valley of the Kings on November 4, 1922.

The contents of King Tut's tomb reflect animal symbolism of ancient Egypt. A perfume vessel with hieroglyphic tadpoles, symbolizing 100,000 years, represents hope for a long life for the pharaoh. The roof of the second outermost shrine is decorated with images of Nekhbet, the vulture goddess and protectress of Egypt and the pharaohs. Two wooden statues depict the pharaoh standing on a black panther. The panther may be an allegorical image of the sky, represented as a feline that swallowed the sun each evening and rejuvenated it the following morning. Two gilded wooden statues depict the king, wearing his crown with the uraeus (upright cobra, symbol of royalty), standing on a papyrus raft while hunting the hippopotamus, symbol of evil. The lion represented a pharaoh's power. Fittingly, two lions' heads look out from the front of the king's throne. The throne's arms are formed from two winged cobras framing the hieroglyphics for "king of Upper and Lower Egypt," and the legs end in lions' paws. Insignias of both the goddesses Nekhbet (vulture) and Wadjet (cobra) appear on the gold funerary mask. The "boy king" was well protected for his journey to the afterlife.

NOVEMBER 5

Animist Independence Day

Followers of animism, the world's most ancient belief system, profess a broad range of spiritual beliefs, but a common denominator is that natural phenomena, including animals, rocks, trees, the wind, etc., have souls or spirits independent of their physical beings. The Latin word *anima* means "soul." Religious experts speculate that our hunter-gatherer ancestors practiced animism. Today, animism has followers all over the world, from Africa to the United States. November 5 is Animist Independence Day, a day animists celebrate humans' rise from ignorance to enlightenment.

In some animist cultures, people revere other animals, often regarding them as relatives, or as residences of the spirits of dead ancestors. For example in parts of Madagascar, people honor crocodiles, believing that after chiefs die, their spirits pass into the reptiles. Some clans believe they are descended from crocodiles. When a person dies, a long nail is driven into the deceased's forehead to keep the corpse immobile. The nail is removed when the body is placed in the family grave a few days later. Once the grave is covered, a tail forms on the body, the hands and feet grow claws, and scales cover the skin. The corpse transforms into a crocodile and joins its ancestors in the river. Belief in the animistic attributes of crocodiles results in people honoring, rather than harming, these potentially dangerous reptiles. Animism can lead to a powerful respect for nature.

Miniature Dinosaurs

Horned lizards, also known as "horny toads," resemble miniature dinosaurs, sporting crowns of horns and sides fringed with large spiny scales. The Navajo honor these lizards as symbols of power, strength, and wisdom, while some other cultures have appreciated horned lizards in less honorable ways. In southern California, from the late 1800s to the early 1900s at least 115,000 horned lizards were stuffed and sold as souvenirs. Countless horned lizards have been kept as pets and have died from improper diet, because most species eat only ants.

Populations of Texas horned lizards have experienced severe declines. In 1990, attorney Bart Cox invited citizens concerned about the declines to a meeting at the Austin Nature Center. Held

on November 6, 1990, and attended by more than 200 people, this meeting gave rise to the Horned Lizard Conservation Society. Two and one-half years later, the society introduced legislation to the Texas State Legislature resolving that the Texas horned lizard be officially designated the State Reptile of Texas. A driving force behind this legislature was the power of kids: 10-year-old Abraham Holland and his younger brother Noah, who convinced their local state representative to support the bill. In June 1993, the bill passed. Calling attention to the horned lizard as the state's reptile gave Texans a greater appreciation for the lizards and increased their willingness to protect them and their habitats.

NOVEMBER 7

Nature's Good Morning

According to legend (no doubt apocryphal), around 750 CE an Ethiopian goat herder named Kaldi watched his goats prance and frolic all night after eating red berries. Monks at a nearby monastery dried the berries and brewed them into a hot drink. They were delighted with the drink, because it allowed them to stay awake through their long hours of prayer and meditation. The red berries were coffee beans. Coffee is now one of the most popular drinks in the world, and more than 75 countries grow coffee.

Kona coffee, cultivated in the Kona district on the Big Island of Hawaii, is one of the most expensive coffees in the world because of high production costs. Missionary Samuel Ruggles planted the first coffee cuttings (from Brazil) in Kona in 1828. By 1841 coffee plantations were established. In 1899, more than 3 million coffee trees grew in Kona. All aspects of production, from planting to picking, are still performed by hand. Many of Kona's coffee growers are fifth-generation farmers. Today, about 650 farms cultivate coffee in the district. Is the high price of Kona coffee worth it? Why not decide for yourself?

Ever since 1970, the 10-day Kona Coffee Cultural Festival has taken place during the coffee harvest season in November, beginning about November 7. The festival includes tours of coffee farms, multicultural music and dance performances, opportunities to pick coffee, and a Kona coffee recipe contest, featuring favorite entrée and dessert concoctions. And, of course, there's lots of coffee to sample.

NOVEMBER 8

Camel Fair

For a unique county fair, check out the five-day Pushkar Camel Fair, held in the state of Rajasthan, India. Each November, more than 300,000 people gather in the dusty desert town of Pushkar to buy and sell camels and other livestock and to enjoy a spectacular festival. The timing of the fair, around the full moon of November, coincides with the Hindu holiday of Kartik Prunima, a time to bathe in Pushkar Lake to be absolved of sin. According to legend, the lake was formed when Lord Brahma dropped a lotus flower from the heavens to crush an earthly demon.

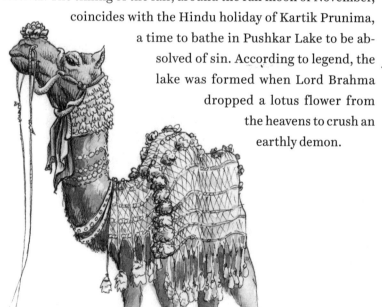

The fair opens with a camel race, and proceeds with folk music, storytelling, dancing, food, magic shows, snake charming, craft bazaars, and camel rides. Competitions include a camel beauty contest, longest waxed mustache, and men's turban-tying. Organic smells of camels, cows, horses, sheep, and goats mingle with enticing odors of spicy chickpea curry, pakora, and samosas. The cries, moans, groans, and bellowing of 50,000 camels and other animals compete with snake charmers' flutes, herders haggling over prices, and musicians entertaining the crowds. Women collect camel dung, which they dry and sell back to the camel herders and traders as fuel for their cooking fires. The fair ends after the full moon, and at dawn thousands of spiritual pilgrims bathe in Pushkar Lake in a purification rite of passage.

NOVEMBER 9

"Ah-ha!"

Carl Sagan was born on November 9, 1934, in Brooklyn, New York. Astronomer, cosmologist, astrobiologist, and astrophysicist, Sagan made it his lifelong quest to understand the universe, especially our planetary system. He was passionate about communicating the thrill of scientific discovery with others, and about making science accessible to the general public. One of his memorable quotes encourages all of us to ask and answer questions about our surroundings: "When you make the finding yourself—even if you're the last person on Earth to see the light—you'll never forget it."

One of the joys of being a scientist is experiencing "ah-ha moments" when things come together. Can you imagine discovering that tides are caused by the moon (Seleucus of Seleucia, 150s BCE), the physics behind rainbow formation (Theodoric of Freiberg, early 1300s), the structure of the cell (Robert Hooke, 1665), that lightning is electrical (Benjamin Franklin, 1751), radioactivity (Henri Becquerel, 1896), the 3-dimensional double helix structure of DNA

(James Watson and Francis Crick, 1953), or that liquid water exists on Mars (Lujendra Ojha and other researchers associated with NASA's Mars Reconnaissance Orbiter, 2015)?

"Ah-ha moments" are not reserved for scientists. Without necessarily realizing it, we all use the inquiry cycle to understand our surroundings. We observe something that makes us ask a question. We gather information to answer the question, and then we reflect on that information. Our reflection leads us to ask another question, and the cycle of inquiry continues.

NOVEMBER 10

The Hope

The ancient Greeks called diamonds *adamas*, meaning "unbreakable." Plato wrote that diamonds were living celestial spirits. The Greeks also held that diamonds were tears of the gods or fragments of stars fallen to Earth. Now we know that diamonds were formed under high pressure and temperature conditions 100 miles (161 km) beneath Earth's surface a billion years ago.

The Hope Diamond, a 45.52-carat blue beauty about the size and shape of a walnut, has a royal history. In 1666, French gem merchant Jean-Baptiste Tavernier bought a crudely cut 112-carat diamond in India. Two years later, he sold the stone to King Louis XIV of France. It was re-cut into a 67-carat stone in 1673 and became known as the "French Blue." In 1749, King Louis XV had the stone reset into an elaborate pendant. After Louis XVI and Marie Antoinette tried (unsuccessfully) to flee Paris in 1791, the crown jewels were turned over to the government. The following year, during the French Revolution, the diamond was stolen. By the time the stone reappeared in 1812, owned by a London diamond merchant, it had been cut to its current smaller size and shape. In the late 1830s, Dutch banker Henry Philip Hope purchased the diamond. It stayed in the Hope family until 1901. After being owned by several others, Washington,

DC, socialite Mrs. Evalyn McLean purchased the Hope Diamond in 1911. Following McLean's death, New York jeweler Harry Winston bought the diamond in 1949. On November 10, 1958, Winston donated it to the Smithsonian Institution in Washington, DC. Since then, more than 100 million museum visitors have viewed the Hope Diamond.

NOVEMBER 11

Flower of Remembrance

> In Flanders fields the poppies blow
> Between the crosses, row on row,
> That mark our place; and in the sky
> The larks, still bravely singing, fly
> Scarce heard amid the guns below.
> —JOHN MCCRAE, "In Flanders Fields"

World War I formally ended at the 11th hour of the 11th day of the 11th month in 1918, with the armistice signed between representatives of Germany and the Allied Nations. The day has been celebrated in the United States ever since, first known as Armistice Day. In 1954, the name was changed to Veterans Day to celebrate the service of all US military veterans and especially to honor war veterans. Veterans Day coincides with Armistice Day and Remembrance Day (also called Poppy Day) in other parts of the world.

During WWI, after soldiers were buried on Flanders' battlefields in western Belgium, dormant poppy seeds germinated and covered the graves. Colonel John McCrae, a surgeon with Canada's First Brigade Artillery, wrote "In Flanders Fields" in early May 1915, expressing his grief over the dead and buried on these battlefields. Vibrant red poppies soon became known as "the flower of remembrance." Since 1921, artificial poppies have been sold on and around November 11 in many countries, including the United States, Brit-

ain, France, Canada, Australia, and New Zealand, to benefit veterans and their loved ones. For many, the poppy serves as a symbol for blood spilled in war. Wearing a poppy is a way of saying thank you to the men and women who have sacrificed so that we might be free.

NOVEMBER 12

Mums

Chrysanthemums ("mums"), a member of the daisy family, are the birth flower for people born in November. Early drawings of chrysanthemums depict them as yellow daisy-like flowers. Many chrysanthemums cultivated today are showy, including some with pompom-like flowers.

Chrysanthemums were first cultivated in China at least 2500 years ago, where they represented life itself because of their perceived medicinal value in treating a variety of ailments and preventing aging. They were considered so powerful that only nobility were allowed to plant them. Today the flowers symbolize powerful *yang* energy and are believed to boost the immune system. Following ancient Chinese tradition, people place a single chrysanthemum petal

in the bottom of a wine glass in hopes of securing a long and healthy life.

During the eighth century CE, chrysanthemums were introduced into Japan. The 16-petal chrysanthemum is the Imperial Seal of Japan, and Japanese emperors sit upon the Chrysanthemum Throne. The golden chrysanthemum appears on the cover of the Japanese passport and on the 50-yen coin. Celebrated on the ninth day of the ninth month, the Japanese Festival of Happiness celebrates the chrysanthemum.

Today, mums are one of the world's most popular flowers. Consistent with the language of flowers from the Victorian Era, they still symbolize friendship, joy, love, and compassion—a perfect flower to give to someone born in November.

NOVEMBER 13

Burden of Punishment

> Ah! Well-a-day! What evil looks
> Had I from old and young!
> Instead of the cross, the Albatross
> About my neck was hung.
> —SAMUEL TAYLOR COLERIDGE, "Rime of the Ancient
>   Mariner" (part II, stanza 14)

In the West Indies, the albatross is said to foretell the arrival of ships, no doubt reflecting the fact that both ships and albatrosses retire into harbors when winds at sea become rough. Widespread maritime lore holds that albatrosses are supernatural because they can fly long distances without flapping their wings. Ever since Coleridge's poem, sailors have considered it bad luck to kill an albatross.

Coleridge began to write "Rime of the Ancient Mariner" on November 13, 1787. The poem's narrator, the Mariner, tells that his ship got caught in a foggy ice field. An albatross soared overhead, apparently to steer the sailors through the fog and ice. Instead of being grateful, the Mariner shot and killed the bird. Soon afterward, the wind died, the temperature rose, and the sailors had no drinking water. The sailors blamed the Mariner for their ill fate and hung the dead albatross around his neck as a burden of punishment. All the sailors died, save the Mariner, who was left with his guilt and the never-ending need to tell his story. He always closed his tale with a lesson: "He prayeth best, who loveth best All things both Great and small, For the dear God who loveth us, He made and loveth all" (part VII, stanza 23).

Lyell and Uniformitarianism

Today is the birthday of Charles Lyell, foremost geologist of his day, born in Scotland in 1797. Lyell studied law and was admitted to the bar in 1825, but his real passion was geology. At the age of 33, he published the first volume of *Principles of Geology*. This landmark book expanded on James Hutton's ideas of uniformitarianism, with the take-home message that the forces of geological change that have shaped Earth for millions of years are still molding the planet today. The book shook the prevailing idea that unique catastrophic events, such as Noah's Great Flood, were responsible for shaping Earth's surface.

Many present-day geologists have come full circle and point out that periodic catastrophes *do* shape Earth's history. Perhaps best known is the hypothesis that the dinosaurs died off from the impact of a powerful asteroid or comet, or that a massive bout of volcanism caused their demise. Astronomical factors might exert a profound influence on Earth's long-term geological processes.

Nonetheless, uniformitarianism is still a central tenet of geology. One significant, and lasting, legacy of Charles Lyell was his influence on Charles Darwin. Darwin had to plan carefully what books, supplies, and personal belongings he packed for his voyage on the *HMS Beagle*. One item he took was a copy of Lyell's book, volume 1, published the previous year. Lyell's ideas led Darwin to think of biological evolution as a slow process during which small changes accumulate gradually over long periods of time.

NOVEMBER 15

"Crikey!"

"Crikey," an Australian exclamation of surprise or amazement, was a favorite with Steve Irwin, a passionate conservationist. Irwin en-

tertained and enlightened people about wildlife through his zoo in Queensland, his television series "The Crocodile Hunter," and other film work. After Irwin died in 2006 when a 12-inch (30-cm) stingray barb pierced his heart, November 15 was designated Steve Irwin Day. Irwin was particularly fond of crocodiles, and he was not alone.

Ancient Egyptians worshipped Sobek, Crocodile God and Lord of the Nile. According to myth, Sobek crawled out of the primordial waters and created the Nile from his sweat. He laid eggs on the riverbank, and the eggs hatched into all of creation. Crocodiles were Sobek's representatives on Earth, believed to bring the rain that overflowed the Nile and fertilized the land. To kill one was punishable by death. In honor of the Crocodile God, Egyptians built a city named Sobek on the west bank of the Nile. Greeks later renamed the city Crocodilopolis. Residents of Crocodilopolis revered a live crocodile named Petsuchos, believed to house Sobek's incarnated soul. Petsuchos lived in a lake near the city's temple, where he was hand-fed cake, honey, milk, and wine. After the death of each Pet-suchos, another crocodile was captured and placed in the temple lake. Each Petsuchos was decorated with golden earrings and brace-lets, and each deceased Petsuchos was mummified and buried in a tomb. "Crikey!," as Steve Irwin might have said.

Time Enough?

> The butterfly counts not months but moments, and has time enough.
> —RABINDRANATH TAGORE, Bengali poet and philosopher;
>   *Fireflies*

Nature-lovers worldwide applaud the spectacular long-distance migrations of eastern monarch butterflies to overwintering sites in Mexico. Less well known is that western monarch butterflies migrate to more than 200 coastal grove sites in California, where they overwinter in clusters by the tens of thousands. In the spring, they migrate east across California and into other western states in search of milkweed on which to lay their eggs. The caterpillars eat only milkweed until spinning their silk chrysalises. After about two weeks inside the chrysalis, a royal-looking bright orange and black butterfly with small white spots emerges. Western monarch butterflies produce four generations during spring, summer, and early autumn. Each generation spreads further from the wintering ground, individual butterflies living only two to five weeks during the breeding season. The last generation migrates back to coastal California, with individual butterflies returning to their ancestors' groves.

Since 2000, volunteer citizen scientists have counted the num-bers of western monarch butterflies during a 3-week period around Thanksgiving. Seventeen years' worth of data from the Western

Monarch Thanksgiving Count reveal dramatic declines in population numbers—nearly 90 percent in many sites. Potential causes include loss of milkweed, loss of overwintering groves or reduced quality of the groves, and climate change. We can only hope that Tagore's sentiment also applies to the long-term future of western monarch butterflies.

NOVEMBER 17

Alaska Bald Eagle Festival

Thousands of years ago, glaciers receded from the Chilkat Valley of southeastern Alaska and left behind a porous gravel bed where the Chilkat River now flows. During the summer, water warmed by the sun sinks into the gravel; during the winter the warmed water slowly percolates through the gravel aquifer and keeps about four square miles (10.4 sq km) of the Chilkat River from freezing. This phenomenon allows for a late run of chum and Coho salmon in mid-

November, which attracts up to 3000 bald eagles—the largest concentration of bald eagles anywhere.

Each year in mid-November, people from around the world visit Haines, Alaska, in the Chilkat Valley, to welcome the salmon and eagles at the Alaska Bald Eagle Festival. The event offers something for everyone, from workshops on photographing the eagles to fieldtrips to view bald eagles at the Chilkat Bald Eagle Preserve. Visitors listen to lectures on the natural history of the valley, southeastern Alaskan wildlife, natural history of salmon, and current bald eagle research; learn about avian anatomy and apply the knowledge in a bird-drawing class; visit the American Bald Eagle Foundation's aviaries and diorama and natural history exhibits; meet avian ambassadors while enjoying wine and hors d'oeuvres; indulge at the Chocolate Extravaganza Buffet; and sample moose, elk, bison, and caribou chili at the Fundraiser Chili Cook-off.

NOVEMBER 18

"God of the Air"

Maya and Aztecs viewed the resplendent quetzal as "god of the air," a sacred symbol of goodness and light. The bird was linked with the feathered serpent god Kukulkan (Maya)/Quetzalcoatl (Aztec), a beneficent god associated with sunshine and vegetation. In the ancient Aztec language of Nahuatl, *quetzal* meant "a large, precious green feather." The quetzal is one of the most spectacular of all Central American birds, with its iridescent emerald-green body and blood-red breast. Breeding males grow twin blue-green tail feathers up to 36 inches (0.9 m) long. Maya and Aztec rulers wore headdresses made from quetzal feathers. Because it was forbidden to kill the birds, they were captured, the feathers plucked, and then released.

Today the quetzal is the national bird of Guatemala. The country's emblem, consisting of a shield with two rifles and two swords crossed with a wreath of laurel, a scroll of parchment with the words

"Liberty of September 1821," and a resplendent quetzal, was created on November 18, 1871—50 years after Guatemala attained its independence from Spain. For the Maya, the quetzal also symbolized freedom because it could not survive in captivity. Quetzals call in a series of deep, smooth notes: "keow, kowee, keow, k'loo, keow, k'loo, keeloo." According to Mayan legend, before the Spanish conquest quetzals sang the most beautiful song of any bird. Ever since the conquest they have hummed mournfully. When the land and the people are truly free, quetzals will sing joyously once again.

NOVEMBER 19

A Burbling Hoot

If you've never heard sandhills, imagine a Scotsman clearing his throat in a long, burred *garooo*! into a rain barrel. Depending on your fancy, this burbling hoot can sound hopeful, mournful, weird, archaic, or simply beautiful. Had the producers of early

dinosaur films possessed a little more imagination, they would
have given their rubber reptiles the voices of sandhill cranes.

—STEVE GROOMS, *The Cry of the Sandhill Crane*

Humans have long revered cranes, admired for their gregarious
behavior, beauty, and graceful long legs and necks. After we die,
according to Japanese, Vietnamese, and other Asian folklore,
cranes carry our souls to Heaven on their backs. We associate cranes
with honorable attributes, reflected by the birds' biology: fidelity
and love because they mate for life; compassion, honor, and loyalty because they excel at maintaining stable family groups; joy and
happiness because they dance exuberantly. We admire cranes for
vigilance because they keep a wary eye cocked as they forage. We
consider the birds to be wise, because their long lifespans provide
considerable time to accumulate knowledge.

Festival of the Cranes is an annual event held the third week
in November at Bosque del Apache National Wildlife Refuge, New
Mexico. The focus is on sandhill cranes, still abundant throughout
much of their range in North America. The spectacle of a large flock
of sandhills and the sound of their primeval cries are unforgettable.

NOVEMBER 20

Princess, No. 799

Thousands of webcams set up all over the world record the daily
activities of animals, from ants to zebras. With a click of the computer mouse, you can become a voyeur and watch animals fight,
play, eat, mate, and care for their offspring in real time. In addition to providing critical data for biologists monitoring and sur-
veying wildlife, these webcams entertain and delight nature-lovers
and allow non-biologists to engage with wildlife in a way previously
unimaginable. They allow us to form emotional connections with
other animals.

The first webcam to stream a condor nest in the wild was set up at the Ventana Wildlife Society Condor Sanctuary in Big Sur, California. On November 20, 2015, a wild California condor dubbed "Princess" fledged under the watchful gaze of faithful followers who had watched the chick hatch, grow feathers, gulp food regurgitated by her parents "Kingpin" and "Redwood Queen," and finally hop about and test her wings. Princess was given a green wing tag with the number 799. The following summer, July 2016, she was seen flying over mountains and valleys and exploring the Big Sur coastline. For the people who watched Princess play her part in helping to bring California condors back from the brink of extinction, the webcams were better than any soap opera.

NOVEMBER 21

Celebrate Those Who Fish

November 21 is World Fisheries Day, celebrated annually since 1998 by fishing communities around the world. From the viewpoint of fisheries, the word fish refers to any aquatic animal that is harvested, from mollusks, echinoderms, and crustaceans to fishes to whales—animals that provide about 25 percent of the world's dietary animal protein. Fisheries touch us in many ways, from providing livelihoods for more than 43 million people to providing food for billions of us.

Just as farmers are connected to the land, so fishers are connected to water. People living in fishing communities are bound together by a shared sense of identity, economic base, and value systems focused on fishing. Their lifestyles and cultures, including homes, cuisine, art, music, literature, folklore, and religious beliefs, are centered on water and aquatic animals. Their daily and seasonal social and work activities and gender roles are structured by the activity cycles of the marine or freshwater life they harvest. Without those who harvest "fish," we would lose a major source of animal protein.

Why not celebrate the day by supporting fisheries? Enjoy some sustainably harvested "fish" in oyster stew, shellfish gumbo, shrimp creole, broiled lobster tails, sea urchin bruschetta or risotto, trout almondine, or grilled salmon. And give thanks to fishers worldwide.

NOVEMBER 22

They Cannot Speak for Themselves

> Until he extends his circle of compassion to include all living things, man will not himself find peace.
> —ALBERT SCHWEITZER

"Reverence for life" was the keystone of Albert Schweitzer's philosophy. And for this, the good missionary doctor received the 1952 Nobel Peace Prize. Buoyed by the worldwide spotlight on Schweitzer's views, the arguments and concerns of animal advocates in the 1950s received increasing attention. It was time to form a new organization that would focus on developing national policy regarding animal welfare. On November 22, 1954, the Humane Society of the United States (HSUS; then called National Humane Society) was founded in Washington, DC.

The HSUS seeks to prevent all activities that cause pain, suffering, or fear to animals. Over the years, the organization has addressed such issues as methods used at slaughterhouses, regulation of animal experimentation, and humane euthanasia of animals in shelters. More recently, the organization has focused on cruelty to animals in industrial agriculture, animal fighting, the fur trade (animal trapping), puppy mills, greyhound racing, and wildlife abuse.

British primatologist Jane Goodall eloquently expressed the need for people to be animal advocates: "The least I can do is speak out for the hundreds of chimpanzees who, right now, sit hunched, miserable and without hope, staring out with dead eyes from their metal prisons. They cannot speak for themselves."

NOVEMBER 23

Nature Spirits

The Japanese religion of Shinto (*kami-no-michi*) focuses on respect for manifestations of nature referred to as *kami*, loosely translated as gods or spirits. These nature spirits, including those of beings (e.g., human ancestors and other animals), elements of the land-scape (e.g., mountains and waterfalls), and forces of nature (e.g., the wind and thunder), live in or often visit natural areas. A prin-cipal *kami*, Inari, the spirit of foxes, fertility, rice, tea, sake, agri-culture, and prosperity, has been revered since at least 711 CE. Of the approximately 80,000 Shinto shrines in Japan, about 32,000 are dedicated to Inari. Inari's messengers are spirits of white foxes. Fox statues often welcome visitors at the entrance to Inari shrines. Offerings of rice, sake, and a sushi roll of fried tofu (*Inari-zushi*) placate the foxes, which are asked to plead with Inari on behalf of worshippers. On Kyushu, Japan's third largest island, Inari Festival Day begins five days before the full moon in November.

Kami play a valuable role in protecting landscapes. Shinto followers have long believed that if the spirits' natural areas are disturbed, the *kami* are likely to retaliate, whereas respectful behavior can yield protection and blessings. Thus, many natural areas, including small forest "islands" in urban areas, have remained undisturbed for hundreds of years even though they are accessible and could be easily exploited. Unfortunately, traditional beliefs in *kami* have eroded in recent years. Some Japanese conservationists advocate educational outreach to revive traditional *kami* beliefs, which would strengthen appreciation of nature.

NOVEMBER 24

On the Origin of Species

During the first half of the nineteenth century, some scientists believed that species change through time—they evolve. Others believed that species are immutable. After Charles Darwin returned from his voyage on the *HMS Beagle* in 1836, he reflected on his observations and asked himself how species might evolve. He hypothesized that individuals better suited to their environment survive longer and produce more offspring, the idea now known as natural selection. Darwin worried that he would be ostracized if he made his idea public. He resolved to gather more evidence while he focused on other writings.

In February 1858, Darwin received a letter from British naturalist Alfred Russel Wallace. Inspired by Darwin's voyage, Wallace had explored the Amazon Basin. Wallace sent Darwin a paper discussing his idea of evolution by natural selection and asked for advice on how to publish it. The two men had formulated the same idea independently! Darwin realized he was about to be scooped. Unable to contact Wallace, who was in the Malay Archipelago, Darwin consulted his colleagues geologist Charles Lyell and botanist Robert Hooker for advice. Lyell and Hooker arranged for a joint 18-page paper to be read by the secretary of the Linnean Society at their meeting on July 1, 1858. Neither Wallace nor Darwin attended. Wallace was still exploring Malaysia, and Darwin was at home mourning the recent death from scarlet fever of his 18-month-old son. The following month, Wallace's formal paper and an excerpt from Darwin's manuscript were published. Darwin knew he'd better get cracking, so he focused on his book and published *On the Origin of Species* on November 24, 1859. The theory of evolution by natural selection is the foundation of modern biology.

NOVEMBER 25

The Wonder of Nature

> What is commonly called "conservation" will not work in the long run, because it is not really conservation at all but rather, disguised by its elaborate scheming, only a more knowledge-able variation on the old idea of a world for man's use only.
> —JOSEPH WOOD KRUTCH, "Conservation Is Not Enough"
> (1954)

American writer, cultural commentator, theater critic, literature professor, and naturalist Joseph Wood Krutch was born on November 25, 1893. During his lifetime he experienced a personal, philosophical, and intellectual odyssey regarding science and nature. In 1929 he published *The Modern Temper*, in which he deplored science for causing skepticism that displaces religious certainty; diminishes or destroys our feelings, emotions, and desires; and leads to a meaningless world. Later, through studying the writings of Henry David Thoreau, Krutch came to believe that humans could discover deep spiritual satisfaction through nature. In 1952, he moved from New York to Tucson, Arizona, and spent the rest of his life celebrating nature by writing about ecological issues and the deserts of the American Southwest.

Joseph Krutch, Aldo Leopold, Archie Carr, and some other early conservationists during the mid-1900s shared a belief that we must discard the idea of human-centered conservation and instead realize that the real worth of wilderness, plants, and animals is in our wonder, admiration, and love of our surroundings.

NOVEMBER 26

Rain

> I think rain is as necessary to the mind as to vegetation.
> My very thoughts become thirsty, and crave the moisture.
> —JOHN BURROUGHS

The world's record rainfall in one minute fell on November 26, 1970—1.5 inches (38 mm), in Barot, Guadeloupe, in the Caribbean. That amount of rain is what some call a gully-washer, frog-strangler, toad-choker, or duck-drownder.

When rain pounding on your roof keeps you awake, think about what it does for us. Rain fills aquifers, lakes, and rivers; provides fresh water for aquatic and terrestrial animals; nourishes plants; and cools and humidifies the air. Rain resurrects life, from dormant seeds to dehydrated tardigrades. Think about the cultural significance of rain, expressed in folklore and mythology throughout history. Ancient Romans worshipped Jupiter, god of sky and thunder, who brought the rain. Ancient Egyptians worshipped the lion-headed rain goddess, Tefnut. The Aztecs sacrificed young children to nourish and propitiate their rain god, Tlaloc. In Hindu mythology, cobras controlled the rain clouds. In Chinese, Japanese, and Vietnamese mythology, dragons controlled water and rainfall. For Aboriginal Australians, Rainbow Serpent brought the rain and nourished Earth. Cultures from Africa to North America perform rain dances to ask the gods for rain. Rather than trust the gods, some cultures seed clouds with chemicals to produce rain. Many people find the sounds, sights, and smells of rain soothing. Rain refreshes, and it moistens our thoughts.

NOVEMBER 27

Queen Conch

Picture yourself on the white sand beach at Providenciales Island, Turks & Caicos, in the Caribbean. In front of you is a smorgasbord of conch: chowder, salad, ravioli, crepes, fritters, empanadas, wontons, sweet'n' sour conch, and pecan-encrusted conch. Word has it that dessert is conch sautéed in rum-and-butter sauce. You have come for the annual Turks & Caicos Conch Festival, held the last Saturday in November. The festival celebrates the islands' national symbol and main export, a marine gastropod: the queen conch. Conch au naturel is mild, sweet, and chewy. Some consider it heavenly.

After your feast, you can participate in the conch-blowing tournament and join a long line of peoples who have made music with conchs, often within a religious or ceremonial context. Folklore dating back to the ancient Greeks claims that, whereas mermaids generally lure men with their enchanting songs, mermen more often make music with conch shells. Aztecs associated the haunting and mellow sounds from conch shell trumpets with the underworld, the moon, and fertility; both priests and warriors used the trumpets. Mayan hunters blew into conch shells to announce their return with slain deer. The Moche of ancient Peru worshipped the sea, and they likewise used conchs as ritual trumpets. Queen conchs are still played, now as musical instruments, in some parts of the West Indies and Caribbean.

Feast and Food Fights

The last Saturday of November sees the Monkey Buffet Festival in Lopburi, Thailand, north of Bangkok. This unique festival honors the 3000 or so long-tailed macaque monkeys that guard the city's ruins and temples built in the tenth century. A local belief holds that these macaques are descendants of the Hindu monkey god, Hanuman. On this day, over 8800 pounds (4000 kilograms) of vegetables, tropical fruits, peanuts, boiled eggs, cakes, and candies are set on long tables covered with red tablecloths in front of the ancient temples for a feast that is sure to please these primates.

A local hotelier named Yongyuth Kitwattananusont initiated the festival in 1989 as a gimmick to boost tourism in the area. It worked. Now, thousands of visitors come every year to watch the monkeys stuff their faces. Once satisfied during the feeding frenzy, the monkeys throw the leftovers at each other in food fights, another crowd-pleaser. Plenty of other activities entice human spectators to the celebration, including music and dancing while wearing monkey costumes.

A word of warning should you attend: The monkeys have become accustomed to people, and they won't hesitate to grab food you might be eating, your camera, cell phone, purse, or other object. When you pay your entrance fee, you might be given a stick to ward off overly enthusiastic and greedy monkeys.

NOVEMBER 29

Patron Saint of Ecologists

In a talk entitled "The Historical Roots of Our Ecological Crisis," presented at the 1966 annual meeting of the American Association for the Advancement of Science, historian Lynn White Jr. made the case that St. Francis of Assisi's attitudes toward nature and animals were profoundly at odds with the official view espoused by the medieval Roman Catholic Church at the time. The view disseminated by the Church held that people were separate from nature, free to exploit animals, plants, and minerals for their own advantage. White argued that modern Western society inherited this destructive attitude toward nature from the Church.

In contrast, St. Francis (1181/1182–1226) taught that all creatures are equal and that we should respect other animals and all of nature. White argued that only by accepting the philosophy of St. Francis can we hope to avert today's environmental crises. He ended his talk by proposing that St. Francis be declared the Patron Saint of Ecologists.

Despite the fact that White blamed the teachings of the medieval Catholic Church for harming the environment, on November 29, 1979, Pope John Paul II proclaimed St. Francis "the heavenly Patron of those who promote ecology." The Pope emphasized that humanity is under sacred obligation to respect the natural world and to protect it from harm.

Kirins and Unicorns

Many of us love the world of make-believe. Fantasy stretches our imagination, blurs the lines between magic and reality, and releases us from the present. Those who "believe" in unicorns delighted in the news reported from the Korean Central News Agency in North Korea on November 30, 2012. Archaeologists had reconfirmed the "Unicorn's Lair," the home of a unicorn ridden by King Tongmyong, founder of the kingdom that ruled parts of China and the Korean Peninsula from the third century BCE to the seventh century CE. Unfortunately, a week later a report surfaced that the discovered cave wasn't associated with a unicorn after all, but rather with the King's kirin, a mythological chimera-like beast with a dragon head sprouting a single horn. It was an unfortunate case of mistranslation into English.

Not to worry. We will always have unicorns, even if we haven't found a unicorn's lair. For over 4000 years, people have believed in them. Babylonians worshipped the unicorn. Ancient Greek and Roman writers praised the healing power of the unicorn's horn, believed to be an antidote against poisons and a cure for diseases. Later, during the Middle Ages (ca. 476–1400 CE) and the Renaissance (ca. 1300–1700 CE), narwhal tusk was sold as "unicorn horn," believed to have magical powers and to protect against poisoned drinks. Since the late 1300s, the unicorn has been Scotland's national animal. The Scots, with their long history of myth and legend, are proud of their association with the unicorn, symbolizing purity and innocence, healing power, and life itself. Unicorns are one of children's favorite animals, thanks to books and film. And imagination.

DECEMBER 1

The Nightingale

> The nightingale sang so sweetly that the tears came into the
> emperor's eyes, and then rolled down his cheeks, as her song
> became still more touching and went to everyone's heart.
> —HANS CHRISTIAN ANDERSEN, "The Nightingale"

Danish author Hans Christian Andersen has long delighted children and those young at heart with his more than 160 fairy tales, translated into 125 languages. Andersen published his first book of fairy tales on December 1, 1835.

Andersen wrote during the Romantic Period, a movement from the late 1700s to about 1850 that encouraged appreciation of the beauty and wonder of nature over commercialization. He projected this philosophy in his popular story "The Nightingale." At first the Chinese Emperor and his subjects are enchanted with the nightingale's beautiful melodies. Eventually they become bored with the bird and obsessed with a wind-up nightingale, studded with

diamonds, rubies, and sapphires. The live nightingale is banished from the empire. After breaking, the artificial bird is repaired but is so delicate that it can be played only once each year. Five years later, Death confronts the Emperor. Suddenly the live nightingale appears at the open window. She sings Death away and is once again admired by the Emperor. Without preaching, Hans Christian Andersen used this tale to share his love of nature and his distrust of technology, his preference for the real over the artificial.

DECEMBER 2

Environmental Protection

> No one is an environmentalist by birth. It is only your path, your life, your travels that awaken you.
> —YANN ARTHUS-BERTRAND, French photographer and environmentalist

During the late 1950s and throughout the 1960s, Americans became increasingly concerned about human-caused damage to the environment, including dwindling resources, pesticide poisoning, oil spills, air and water pollution, and solid waste disposal. An environmental movement was born, demanding increased protection of human health and the environment.

On December 2, 1970, President Richard Nixon signed an executive order creating the United States Environmental Protection Agency (EPA) to set, monitor, and enforce national guidelines regarding air and water quality and toxic substances. Early charges of the EPA were to administer the 1970 Clear Air Act (to abate air pollution), the 1972 Federal Environmental Pesticide Control Act (to control the sale and use of pesticides), and the 1972 Clean Water Act (to regulate municipal and industrial wastewater discharges). Since then, the EPA has written and enforced regulations regarding vehicle fuel efficiency, oil pollution, ocean dumping, asbestos hazards, uranium mill tailings, protection from radiation exposure, water efficiency, safe drinking water, and climate change issues.

In 2017, the fate of the EPA became uncertain. The Trump administration slashed the agency's budget, reduced staffing, eliminated programs, barred new contracts and grants from being awarded, and instructed the EPA to delete the climate change page from its website. What will happen with reduced environmental protection?

DECEMBER 3

An Immense Intellectual Anthill

> Scientists at work have the look of creatures following genetic instructions; they seem to be under the influence of a deeply placed human instinct. They are, despite their efforts at dignity, rather like young animals engaged in savage play. When they are near to an answer their hair stands on end, they sweat, they are awash in their own adrenalin. To grab the answer, and grab it first, is for them a more powerful drive than feeding or breeding or protecting themselves against the elements.
>
> —LEWIS THOMAS, "Natural Science," in *The Lives of a Cell*

Lewis Thomas (1913–1993) was an American physician, researcher, policy advisor, poet, and essayist. Thomas' first book, *The Lives of a Cell: Notes of a Biology Watcher*, contains 29 essays written between 1971 and 1973 for the *New England Journal of Medicine*. The essays range from biology and anthropology to music, medicine, and mass communication, with an underlying theme of the interconnected web of life.

In his essay "Natural Science," Thomas explains why scientists seem so peculiar. He suggests that scientists have the compulsion to discover written into their genes. There's no stopping, for the answer to one question requires an answer to the next. Many people assume science to be a lonely endeavor, but in fact it is quite the opposite. Thomas suggests that scientists are interdependent, "like an immense intellectual anthill." Thomas died on December 3, 1993. His loss from the metaphorical ant colony is a loss not only for his fellow scientists, but for all society, for he was truly able to translate the mysteries of biology to non–colony members.

DECEMBER 4

World Wildlife Conservation Day

One hundred eighty-three countries have signed the Convention on International Trade in Endangered Species of Wild Fauna and Flora (CITES), which forbids the selling of endangered species to another country. Nonetheless, considerable illegal trade occurs on the black market, including commercial wildlife trafficking of leopard skins and elephant tusks destined to become souvenirs; bear gallbladders, pangolin scales, tiger bones, and rhinoceros horns for traditional medicines; parrots, marmosets, and chameleons for the pet trade; and peccaries, lemurs, great white sharks, and turtles sold as meat. Animals and their parts are smuggled in the same ways as illegal drugs—in purses, suitcases, crates, and on the human body.

In 2012, Secretary of State Hillary Clinton issued a call for action by declaring December 4 as World Wildlife Conservation Day, a day to inform the public about illegal trade in wildlife. Clinton declared, "Wildlife cannot be manufactured. And once it's gone, it cannot be replenished. Those who profit from it [wildlife trafficking] illegally are not just undermining our borders and our economies, they are truly stealing from the next generation."

In response to the growing concern about illegal wildlife trafficking, the US Department of State encourages people worldwide to sign a wildlife pledge (http://www.wildlifepledge.org/). On July 22, 2016, I signed, and at that time 9533 other individuals had pledged to protect and respect the world's wildlife. Please consider taking this important pledge yourself. Do your part to stop wildlife crime by promising never to buy illegal wildlife products.

DECEMBER 5

Hornbill Festival

> An upstart is a sparrow eager to marry a hornbill.
> —MaLAWIAN PROVERB

Hornbills earned the moniker for their large, down-curved, heavy bills used for constructing nests in tree cavities, preening, plucking fruit from branches, and snatching insects, frogs, small reptiles, and mice. Nearly 60 species of hornbills live in tropical and subtropical Africa and southern Asia. Some can exceed 3 feet (0.9 m) in length. Most species are monogamous, and many mate for life.

The birds' majestic, powerful appearance and monogamous behavior engender our admiration. For people of the Indonesian island of Sumba, the Sumba hornbill symbolizes fidelity. For the Poro of the Ivory Coast, the yellow-casqued hornbill symbolizes wisdom and authority. The Iban of Borneo believe the rhinoceros hornbill

relays their messages to spirits of the upper-world; for the Dyak, the rhinoceros hornbill brings good fortune. Various cultures revere hornbills for their perceived protection against evil spirits, lightning, drought, and food shortage.

The great Indian hornbill from Nagaland, northeast India, is associated with beauty and bravery, symbolizes reproductive and agricultural fertility, and is featured prominently in dances and songs performed during an annual festival. All 16 major tribes of Nagaland assemble in early December for a 10-day Hornbill Festival to celebrate the diverse Naga cultural traditions through performing arts, games, sports, and craft and food fairs. From Africa to India to Indonesia, hornbills are appreciated by people as well as by sparrows.

DECEMBER 6

Supatá Golden Frog Festival

With an estimated 32 percent of frog species threatened with extinction, discovery of a new species is exciting. In 2007, scientists discovered a beautiful poison frog in fragments of Andean cloud forest in the municipalities of Supatá and neighboring San Francisco, Colombia. The front half of the body and upper forearms of these small frogs (0.7 inches; 18 mm) is golden-yellow; the rest of the body is sepia-brown with large pale blue or bluish-green spots. These frogs court and fertilize their eggs in damp leaf litter on the ground. After the eggs hatch, the male carries the tadpoles on his back and deposits them in water-filled tanks of bromeliads. Named the Supatá golden frog (*Andinobates supatae*), they are known only from this area, which has already experienced considerable habitat destruction.

Residents of Supatá have honored the frogs by declaring them a natural heritage treasure and their adopted ambassador. Supatá boasts a large statue of the frog in its town square, and ever since

2008 an annual Supatá Golden Frog Festival has been held in early December. The festival focuses on community education, awareness, and conservation of the frog and its threatened forest habitat. Music, dancing, food, and a race to raise funds for reestablishing areas of suitable habitat provide an unusual festival with frogs as its centerpiece.

DECEMBER 7

Protection, Decoration, and Medication

Holly is one of the birth flowers for December. Human cultures have a long and varied association with holly. The ancient Celts planted holly outside their homes to ward off evil spirits and lightning. They brought holly branches inside their homes during winter to secure good luck, and to invite the woodland spirits inside as a safe haven from the cold. Ancient Romans decorated with garlands of holly for their midwinter feast, Saturnalia. Early Roman Christians used hol-

ly as a Christmas decoration, a tradition perhaps derived from the pagan use of holly during Saturnalia. One Christian legend tells that because the cross was made from holly wood, the tree has been reduced to a shrub with thorny leaves. Another legend says that originally holly berries were yellow, but after holly formed the Crown of Thorns, they became red from Christ's blood.

Holly has long been used to treat colic, fever, rheumatism, asthma, and gout. An old English remedy for intestinal worms prescribed that a holly leaf and bits of sage be placed in a dish of water. The person should lower his or her head over the dish, yawn, and the worms would crawl into the person's mouth and drop into the water. No doubt more effective, the poisonous berries were used as an emetic. However you December babies want to use holly—as protection, a decoration, or as a medicament—enjoy your birth flower.

DECEMBER 8

An Ornithologist's Tool

In about 1595, a peregrine falcon belonging to King Henry IV of France disappeared while pursuing a bustard. It was found 24 hours later in Malta, 1350 miles (2173 km) away, having flown about 56 miles per hour (90 km/hr). How do we know it was the same bird? A metal band had been placed around its leg for identification in case it was ever lost. Until that time, no one had any idea that falcons flew that far and with that speed. Fast-forward to 1803, North America, and John James Audubon, who tied silver cords to the legs of nestling phoebes. He identified two of the birds when they returned to the neighborhood the following year. In 1899, Danish schoolteacher Hans Mortensen placed handmade aluminum rings on the legs of European teals, pintails, white storks, gulls, and several types of hawks. He stamped each band with his name and address so that, if found, the birds could be returned. Mortensen's system of banding became the model for our current banding efforts. Three years later,

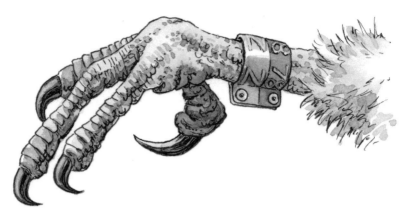

Paul Bartsch of the Smithsonian Institution began the first scientific system of bird-banding in North America by banding more than 100 black-crowned night herons. Not long afterward, on December 8, 1909, Leon Cole of the University of Wisconsin founded the American Bird Banding Society.

Why band birds? A small metal or plastic band with a unique number around a bird's leg allows for future identification, enabling scientists to study population size and growth, migration, dispersal, behavior, life span, survival rate, and reproductive success. The information gathered can be vital for protecting birds.

DECEMBER 9

The Midwife Toad Incident

Gladwyn Kinsley Noble, curator of herpetology and experimental biology at the American Museum of Natural History, died at the age of 46 on December 9, 1940. He will forever be associated with exposing the infamous "midwife toad incident" and Paul Kammerer, an Austrian biologist accused of falsifying his data.

Instead of evolution by natural selection, Kammerer endorsed the theory of French naturalist Jean-Baptiste Lamarck that organisms pass to their offspring characteristics they have acquired during their lifetimes. Kammerer worked hard to prove this theory with elaborate experiments, one of which involved European mid-wife toads (*Alytes obstetricans*). These toads deposit and fertilize strings of eggs on land; the male maneuvers the egg strings onto his hind legs and carries them on his body. He dips into water when the eggs are about to hatch, and the tadpoles swim off. In his experiment, Kammerer forced the toads to lay their eggs in water. He reported that, as is the case with many aquatic-breeding frogs, the males developed black nuptial pads (epithelial swellings) on the thumbs of their front feet to securely hold onto females during the underwater mating, and that this characteristic was inherited by the next generation.

Noble doubted the results. He examined the black patches microscopically, and revealed in the scientific journal *Nature* that the areas had been injected with India ink. Kammerer blamed the alterations on his laboratory assistant. Six weeks after the accusation in 1926, Kammerer committed suicide, at age 46. It may never be known whether Kammerer falsified his data or if the lab assistant sabotaged the experiment. Either way, midwife toads do not develop nuptial pads when forced to breed in water, and Lamarck's theory remains discredited.

DECEMBER 10

Other Animals Have Rights Too

On December 10, 1948, the General Assembly of the United Nations proclaimed the Universal Declaration of Human Rights to prevent atrocities like those that occurred during WWII from ever happening again. The UN declared that respect and dignity are "the foundations for freedom, justice and peace in the world." In 1950, the UN passed a resolution making December 10 Human Rights Day. Later, animal rights activists proposed that the rights of respect and dignity be extended to nonhuman animals, and that December 10 also be celebrated as International Animal Rights Day.

A Gallup poll released in May 2015 reveals that Americans' attitudes toward animals are improving. In 2003, 25 percent of people polled said that animals deserve the same protection from exploitation as we do; that figure has now risen to 32 percent. In 2015, the poll asked whether respondents were concerned about the treatment of animals in different settings. Over 65 percent of respondents are worried about animals used in circuses, sports, and research. More than 50 percent are worried about the treatment of animals in aquariums and zoos and of livestock raised for food. In response to public opinion, countries are banning the use of wild animals in circuses. The US is retiring its federally owned research chimpanzees, and countries across the globe are banning use of animals for testing cosmetics. These changes suggest that we are increasingly recognizing and honoring the rights of other species.

DECEMBER 11

International Mountain Day

> Climb the mountains and get their good tidings. Nature's peace will flow into you as sunshine flows into trees. The winds will blow their own freshness into you, and the storms their energy, while cares will drop away from you like the leaves of Autumn.
> —JOHN MUIR, *The Mountains of California*

John Muir loved the Sierra Nevada Mountains of California and passionately fought for their preservation. His words written in a letter to his sister Sarah say it all: "The mountains are calling and I must go, and I will work on while I can, studying incessantly."

In 2002, the UN General Assembly designated December 11 as International Mountain Day, to raise awareness about the importance of mountains. Covering 22 percent of Earth's land surface, mountains are home to 13 percent of the human population, house high biodiversity, and provide water for at least half of the world's

people. Mountains hold spiritual meaning for many cultures and religions, for example Mount Olympus (Greece—Greek gods); Mount Sinai (Egypt—Judaism, Christianity, Islam); Mount Kailash (Tibet—Buddhism, Bon, Hinduism, Jainism); Mount Fuji (Japan—Buddhism, Shintoism); Croagh Patrick (Ireland—Paganism, Catholicism); Machu Picchu (Peru—Inca); and the San Francisco Peaks (US, Flagstaff, Arizona—Navajo, Havasupai, Hopi, Zuni). Mountains symbolize endurance, strength, constancy, and eternity. They beckon us to climb to the top, to challenge ourselves.

DECEMBER 12

National Poinsettia Day

In 1825, Joel Roberts Poinsett, an amateur botanist and first US Minister to Mexico, sent cuttings of an unusual red and green shrub from southern Mexico to his home in South Carolina. By 1836 the plant had become widely known as the poinsettia, but the flower's story in the United States had just begun. In 1900, a German emigrant named Albert Ecke settled in Los Angeles, where he opened a dairy farm and orchard and sold field-grown poinsettias from roadside stands. In the 1920s, his son Paul Ecke developed a grafting technique for poinsettias, making them fuller and bushier. Later,

Paul Ecke Jr. made the plants more accessible by expediting shipment. Instead of shipping mature plants by railroad, he shipped cuttings to growers by air.

In 2002, the US House of Representatives created National Poinsettia Day to be celebrated on this day to mark the December 12, 1851, passing of Joel Poinsett, and to honor the "father of the US poinsettia industry," Paul Ecke Jr.

Association between poinsettias and the Christmas season dates to a legend from sixteenth-century Mexico. A poor girl named Pepita worried she had no worthy gift to celebrate Jesus's birthday. An angel told her that any gift given with love was worthy. Pepita gathered a bouquet of scraggly weeds and placed them on the church altar. Miraculously, the weeds turned into beautiful red, star-shaped flowers. For Christians and non-Christians alike, the poinsettia is a festive part of the winter holiday season.

DECEMBER 13

Driven to Extinction

Chinese legend tells that a beautiful young girl lived with her evil stepfather on the banks of the Yangtze River. One day, while in a boat on the river, the stepfather tried to take advantage of her. Panicked, she dove into the water and drowned. A large storm battered and sank the boat. After the storm cleared, villagers saw a graceful dolphin swimming in the river—the incarnation of the young girl. The Chinese river dolphin became known as "the Goddess of the Yangtze."

Folklore regarding these pale blue-gray dolphins, known as *baiji*, dates back at least to 200 BCE. Another legend tells that the Goddess of the Yangtze is the incarnation of a beautiful princess, drowned by her family for refusing to marry a man she did not love.

Population size of the Chinese river dolphin, found only in the Yangtze River, was estimated to be 6000 in the 1950s. By the late

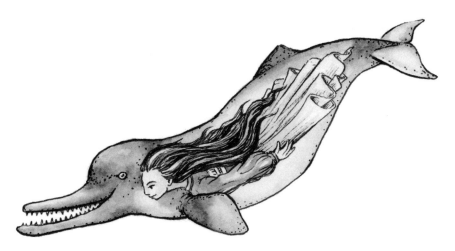

1980s the number had dropped to 400, and by 1997 only 13 sightings were made. The last documented sighting was recorded in 2002, and on December 13, 2006, the Chinese river dolphin was declared extinct. The Yangtze River is one of the world's busiest waterways. Fisheries, pollution, dam-building, river traffic, and poaching have caused the dolphin's extinction. Local fishers have long regarded the Chinese river dolphin as their goddess of protection. What will they do, now that the goddess is gone?

DECEMBER 14

Respect for Primates

Today is World Monkey Day, celebrated on December 14 since 2000. The day honors all primates, not just monkeys. We relate to monkeys, apes, lemurs, and other primates because we are one of them. The bond existed long before we understood that these other primates are our closest relatives. Ancient Egyptians kept baboons as pets; they trained them to climb fig and date trees and toss down the fruit. People still keep primates as pets. Ancient Romans used nonhuman primates for experimentation because they share so many similarities with humans. We still experiment with other primate species and observe their behavior to learn about ourselves.

In January 2017, scientists reported that 75 percent of the 505 species of primates are declining; 60 percent are threatened with extinction. The activities of humans—one species of primate—have caused the decline of these other primates. We destroy their habitat through logging, ranching, farming, mining, and drilling for oil. We hunt them for food, pets, and traditional medicines.

With concerted effort we may be able to reverse some declines, as for example we accomplished with the golden lion tamarin from the Atlantic coastal forests of Brazil. The once-common species was nearly exterminated when forests were replaced with agriculture. In 1983, the US National Zoo led an international effort to breed the monkeys in captivity. Some remaining Brazilian forest was preserved, hunting was banned, and captive-bred monkeys were reintroduced. A stable population of about 3500 golden lion tamarins now lives in the wild.

DECEMBER 15

Halcyon Days

According to Greek myth, King Ceyx of Thessaly drowned when his ship sank during a violent storm. His wife, Halcyone (also known as Alcyone), threw herself into the waves in grief. Taking pity on the couple, the Greek gods transformed them into kingfishers and promised they would live forever. Since then, every year in midwinter Halcyone builds her floating nest and broods her chicks on the Aegean Sea. Halcyone's father, Aeolus, god of the winds, calms the seas and ensures that the winds and waves do not disturb the nest for the next 15 days. This period of placid weather around the winter solstice became known as the Halcyon Days, a safe time for sailors to travel between winter storms. Traditionally, December 15 is considered the first of the Halcyon Days.

This classical Greek myth, recorded by the Roman poet Ovid (43 BCE—17/18 CE), has influenced the scientific nomenclature of

these showy birds. The common kingfisher belongs to the genus *Alcedo*, and its common name is halcyon. Tree kingfishers, including the four species of Australian and New Guinean kookaburras, are in the family Halcyonidae. River kingfishers belong to the genus *Ceyx*, and the belted kingfisher is *Megaceryle alcyon*.

Today, "halcyon days" carries a broader meaning. We use halcyon days to refer, often with nostalgia, to some happy time in the past, such as the carefree days of youth.

DECEMBER 16

Goin' Home

> Going' home, goin' home, I'm a goin' home;
> Quiet-like, still some day, I'm jes' goin' home.
> It's not far, jes' close by
> Through an open door;
> Work all done, care laid by,
> Goin' to fear no more.
> —WILLIAMS ARMS FISHER, "Goin' Home"

Czech composer Antonín Dvořák visited the United States and directed the National Conservatory of Music in New York City from 1892 to 1895. During this time, he wrote his two most successful orchestral works: Cello Concerto in B Minor and Symphony No. 9 in E minor, the "New World Symphony." The symphony premiered in New York on December 16, 1893. The audience loved it, and it remains one of the world's most popular symphonies.

The landscape of America, especially the wide-open spaces and stillness of the prairies, inspired the New World Symphony. The piece reflects Dvořák's joy and wonder at his new environment. In Czechoslovakia, Dvořák had woven rhythms from traditional Bohemian music into his compositions. In the United States, he was intrigued with Native American music and African-American spirituals. He incorporated what he called "the spirit" of African-American melodies in the New World Symphony and said that Native American music had influenced part of his symphony. In 1922, one of Dvořák's students, Williams Arms Fisher, wrote words to the second movement, a much-loved song known as "Goin' Home."

Neil Armstrong carried a recording of Dvořák's New World Symphony on the Apollo 11 mission to the moon in July 1969. How fitting that Armstrong, exploring a "new world," chose to take Dvořák's symphony.

DECEMBER 17

We Can Fly

Birds fly. Bats fly. Insects fly. Pterosaurs flew. We yearn to fly also. Ancient Greeks and Romans gave their gods and mythical beasts the power of flight. In one well-known Greek myth, Icarus, with his feathered wings held together with wax, became intoxicated by the thrill of flying and soared too close to the sun. His wings melted, and he fell and drowned in the ocean. And of course there's Pegasus, the pure white, winged stallion. Long ago, people strapped feathers to their arms and leaped from hills, cliffs, and towers—often to their death—in desperate attempts to fly. In the 1480s, Leonardo da Vinci studied bird flight and sketched designs for flying machines.

On December 17, 1903, the Wright brothers made the first controlled, sustained human flight. During the preceding years, Wilbur and Orville had made over 1000 glides. They learned about sustained lift and how to control an aircraft while in flight. Then they built "Flyer," powered by a crude gasoline engine. Wilbur had a failed attempt on December 14 when he climbed too steeply, stalled, and dove into the sand of North Carolina's Outer Banks. Repairs took three days. On December 17, it was Orville's turn. He kept "Flyer" aloft for 12 seconds. The brothers took turns flying several more times that day, the last being an impressive 852 feet (260 m) in 59 seconds. This was the real McCoy—powered flight. We may not have feathered wings of our own, but we have learned how to fly.

DECEMBER 18

Epona, the Horse Goddess

Celtic legend tells that a man named Fulvius Stellus hated women and instead consorted with his favorite mare. The mare gave birth to a beautiful goddess named Epona ("mare goddess" or "she who is like a mare"), protectress of horses, stables, riders, and horse owners. Epona could take the form of a white mare or a woman. As the latter, she was often depicted riding a white mare sidesaddle. People throughout the Celtic world worshipped Epona.

The invading Roman army became so enamored with this Celtic horse goddess that they adopted her cult and spread it throughout the Roman Empire, where she was worshipped at shrines between the first and third centuries CE. Travelers draped Epona statues with garlands of roses. In Gallo-Roman religion, Epona was also a fertility goddess. As such, she was often depicted sitting or standing between two horses or foals and carrying a basket of grain, bread, or fruit, symbolizing Earth's abundance. She was sometimes depicted feeding her horses from a cornucopia filled with corn and apples. Epona was the only Celtic divinity worshipped in Rome itself. Epona's Feast Day, the only Roman celebration to honor a Celtic deity, was held on December 18 in the Roman calendar. Romans celebrated the day by giving their horses a day of rest, and by honoring the goddess who protected their valuable animals.

DECEMBER 19

Feed Me, Seymour!

One day during a solar eclipse, a nerdy young man named Seymour buys an exotic plant from a Chinese florist. He dubs it Audrey II, after his coworker (and love interest) Audrey in Mushnik's Flower Shop. The plant thrives on insects, but in time it droops, close to death. After Seymour accidentally pricks his finger, Audrey II sucks the blood. Now Audrey II must drink human blood to survive. As she grows, she demands more and more. "Feed me, Seymour, feed me!" Seymour upgrades Audrey II's diet from blood to human flesh. I won't spoil the ending for those who don't know the story of "Little

Shop of Horrors." The film adaptation of this hilarious rock horror musical was released on December 19, 1986.

Real-life carnivorous plants such as Venus flytraps, pitcher plants, and sundews provided the model for Audrey II. Certain plants growing in acidic bogs and other nitrogen-poor soils get some or most of their nutrients from trapping, killing, and consuming animals. Nearly 600 species of carnivorous plants are distributed on every continent except Antarctica. Most trap insects, but spiders, earthworms, and even small vertebrates, including fish, frogs, lizards, birds, and rodents, have been found in the "bowels" of carnivorous plants. Perhaps we find carnivorous plants fascinating because of the unexpected novelty. We think of animals as eating plants, not the other way around. How implausible that a pitcher plant traps and then slowly digests its victim, drowned in a pool of liquid!

DECEMBER 20

Singing and Fighting

Charles Dickens published *The Cricket on the Hearth* on December 20, 1845. In this sentimental love story, a cricket chirps from the family's hearth, acting as a guardian angel and symbol of happiness and good luck. Worldwide and throughout time, people have attributed great significance to the cricket.

For at least the past 1400 years, the Chinese have valued crickets both for their singing and fighting abilities. Historical records from the Tang Dynasty (618–907 CE) reveal that imperial concubines kept crickets in tiny gold cages and took them to bed to hear sweet music throughout the night. From that custom evolved the popular hobby of raising crickets, a pastime still enjoyed by millions of Chinese, who cage male crickets and hang them in their homes, enjoying the constant cheer from the crickets' songs. People carry pet crickets in their pockets, housed in tiny bamboo cages.

Others fight their crickets, a popular sport that sometimes involves gambling. Two crickets are weighed and put together in a box. The crickets bite each other, sometimes ripping off each other's legs. Still they fight. The fights usually end in a few minutes, when one cricket dies, but they occasionally last 45 minutes. Champion fighter crickets sell for over $1000 each. In the late 1990s, an estimated 300,000 to 400,000 people participated in cricket fights in Shanghai alone, involving up to 100,000 crickets per day during the summer-autumn peak season. Like too many other animals, crickets are both admired and exploited.

DECEMBER 21

Double-Sided Nature

In September 1620, 102 passengers, many of them Puritan Separatists, sailed on the *Mayflower* from Plymouth, England, toward the New World seeking religious freedom and/or the hope of a better life. They landed at present-day Plymouth, Massachusetts, on December 21, 1620.

A journal entitled *Of Plymouth Plantation*, written by William Bradford, one of the Puritans and long-time governor of Plymouth, reveals that these early Pilgrims felt the profound power of nature. They struggled against her forces, first facing dangers and hardships during their ocean voyage—rough seas, strong north-Atlantic gales, and water leaks in their ship. Their first winter was harsh, nutrition was poor, diseases such as scurvy were rampant, and housing was inadequate. By the end of winter nearly half of the colonists had died. But then, less than a year later they gave thanks for a bountiful harvest of corn, beans, and squash. Squanto, a member of the Patuxet tribe, had taught the colonists how to catch menhaden (fish) and bury them as fertilizer when planting corn and other crops to increase their yield. Plenty of animal protein was available for the catching: deer, duck, goose, fish, and shellfish.

The Pilgrims must have been acutely aware that nature was a force beyond their control, yet also key to their very survival. This snapshot in time reflects our relationship with nature worldwide and throughout time: feared and revered.

DECEMBER 22

Winter Solstice

Winter solstice in the Northern Hemisphere, most often occurring on December 21 or 22, happens when the sun reaches its southernmost position as seen from Earth. (In the Southern Hemisphere, this is the June solstice.) It is the shortest day of the year and the longest night. Cultures worldwide have long celebrated the event with dance, song, and feasts, a time that symbolizes rebirth and new beginnings as the days begin to lengthen.

The lives of Neolithic (New Stone Age) farmers would have been intimately tied to the changing seasons and movements of the sun. Experts speculate that the Neolithic passage tomb of Newgrange in eastern Ireland, built around 3200 BCE, had astrological, spiritual, and ceremonial significance. Newgrange is a grass-covered circular mound, 280 feet (85 m) in diameter, covering 1.1 acres (0.45 hectare). Within this mound of alternating layers of earth and stone lies a 60-foot (18-m) passageway and interior chambers. Newgrange may have been designed for ritualistic capturing of the sun on the shortest day of the year. On the winter solstice, the rising sun shines through a window-like opening, the "roof-box," strategically placed directly above the main entrance. The sun's rays gradually extend along the passageway, and for 17 minutes they pierce the darkness and dramatically illuminate the inner chamber, revealing rock carvings. New Stone Age farmers of long ago may have built Newgrange to rejoice in new beginnings, the same reasons we celebrate the winter solstice today.

DECEMBER 23

A Blue Fin

Perhaps December 23, 1938, started off like any other day for Marjorie Courtenay-Latimer, curator of a tiny museum in East London,
South Africa. But later that day she got a call that her angler friend
had returned with his catch and that she could take anything she
wanted for her museum. At the docks, she noticed a blue fin stick-
ing out from a pile of sharks and rays. Beneath the pile she found a
5-foot (1.5-m) pale mauve-blue fish with iridescent silver markings.
She had no idea what it was. A taxi driver reluctantly drove her and
the reeking fish back to the museum. She mailed a crude sketch of
the animal to Professor J. L. B. Smith at Rhodes University.

Smith immediately identified the fish as a coelacanth and
named it after its discoverer—*Latimeria chalumnae*, the West Indian Ocean coelacanth. The coelacanth was one of the most important
zoological finds of the twentieth century. Why? These lobe-finned
fish, covered in thick scales that function as armor, were thought
to have gone extinct about 65 million years ago, along with the dinosaurs. Until 1938, the only coelacanths known to science were
fossils.

In 1997, Mark Erdmann, a marine biology student from the University of California, Berkeley, while on his honeymoon in Indonesia, saw a brown (not blue) coelacanth in a market on the island of Sulawesi. This turned out to be a second species of coelacanth, the Indonesian coelacanth (*Latimeria menadoensis*). The West Indian Ocean coelacanth is listed as Critically Endangered, with fewer than 500 individuals. The Indonesian coelacanth is listed as Vulnerable, with less than 10,000 individuals. Will a third species be discovered?

DECEMBER 24

Now, Dasher!

Every Christmas Eve, December 24, the flying reindeer team transports Santa and his sleigh to deliver gifts to good little boys and girls. "Now, Dasher! Now, Dancer! Now, Prancer and Vixen! On, Comet! On, Cupid! On, Donner and Blitzen!" In 1821, the first-known written words connecting Santa with flying reindeer were published anonymously in New York in a 16-page booklet: "Old Santeclaus with much delight, His reindeer drives this frosty night. O'er chimneytops, and tracks of snow, To bring his yearly gifts to you." Two years later, the *Troy Sentinel* published the poem we now call "'Twas the Night before Christmas."

How did this image of flying reindeer come about? Some suggest the idea is a spin-off from the Norse and Germanic legend of Thor, god of thunder, who soared through the sky in a chariot pulled by two large, horned goats: Tanngrisnir ("teeth-barer, snarler") and Tanngnjóstr ("teeth-grinder"). Others focus on Lapland (Finland). For many centuries Laplanders have domesticated reindeer for pulling sleds and sleighs. In a parallel world, Lapp shamans falling into trances after ingesting the psychoactive red and white fly agaric mushroom (*Amanita muscaria*) have reported seeing reindeer flying through the sky. Other hallucinating Lapp shamans have

imagined their spirits being carried away in an aerial sleigh led by reindeer. Add a pudgy gift-giver, and you get Santa and his team. Whatever the origin, and whatever your religious beliefs, we can all celebrate the human creativity of the image.

DECEMBER 25

Two Hundred Twenty-Four Birds

> On the first day of Christmas my true love sent to me
> A Partridge in a Pear Tree.
> On the Second day of Christmas my true love sent to me
> Two Turtle Doves
> And a Partridge in a Pear Tree.
> —"The Twelve Days of Christmas"

For Christians, the "12 days of Christmas" represent the 12 days between the birth of Christ and the Epiphany, the celebration of the Magi's ("Three Wise Men") visit to the Christ child. By visiting and offering gifts, the Magi were the first Gentiles to publicly recognize the divinity of Jesus. Worldwide, many Christians celebrate the 12 days by exchanging gifts on all 12 days. The song "The Twelve Days of Christmas," piped through stores and supermarkets for weeks leading up to the Christmas holidays, reflects this tradition. The gifts in the song become increasingly grand, culminating with all the previous gifts plus 12 drummers drumming.

This Christmas carol was published in England in 1780, without music, in a book intended for children, but it is believed to be French in origin. The song likely was used as a children's memory game. Many of the gifts are birds: 1 partridge in a pear tree, 2 turtle doves, 3 French hens, 4 calling birds, 6 geese a-laying, and 7 swans a-swimming. The 5 golden rings might also refer to birds—to the yellowish rings around the necks of ring-necked pheasants. That makes 224 birds the true love gave to his/her sweetheart. Such was the value of birds during the eighteenth century when the words were written.

DECEMBER 26

Count, Not Kill

During the 1800s, hunters in the United States enjoyed a holiday tradition known as the Christmas "Side Hunt." They teamed up, chose sides in a field, spread out, and shot as many birds as they could. The team of hunters that killed the most birds won the competition. On Christmas Day in 1900, ornithologist Frank M. Chapman proposed a new holiday tradition: counting instead of killing birds. On that Christmas Day, 27 birdwatchers in the US and Canada counted about 90 species.

Now, tens of thousands of birdwatchers across the US, Canada, and other countries sign up each November to participate in the Audubon Christmas Bird Count, which takes place from December 14 to January 5. The 117-year-old event is the world's longest-running citizen-science project. A team of at least ten observers, one of whom is the compiler, counts every bird seen within a circular area (15 mile/24 km diameter) on a given day. The numbers of birds per species are then tallied. In 2013, the record number of species to date for a designated area was recorded: a site on the eastern slope of the Andes in Ecuador yielded 529 species.

Citizen-science volunteers who participate in the Audubon Christmas Bird Count provide data for biologists studying the status and range of birds, including winter survival rates and migration patterns. And, of course, the volunteers have a lot of fun doing what they love to do—bird-watching.

DECEMBER 27

Emblem of Vigilance

Rattlesnakes, unique to the Americas, were the first animals chosen to symbolize the colonies that would become the United States of America. Benjamin Franklin initiated this association, in 1751.

After fighting had begun between the colonists and the British, but before the Declaration of Independence, Benjamin Franklin wrote that he considered the rattlesnake the perfect symbol for the American spirit. In a letter published in the *Pennsylvania Journal* on December 27, 1775, Franklin wrote, under the pseudonym An American Guesser: "I recollected that her eye excelled in brightness, that of any other animal, and that she has no eye-lids. She may therefore be esteemed an emblem of vigilance. She never begins an attack, nor, when once engaged, ever surrenders. She is therefore an emblem of magnanimity and true courage."

The rattlesnake Franklin described in his letter had 13 rattles, firmly united together—exactly the number of colonies united in America. In his words, "'Tis curious and amazing to observe how distinct and independent of each other the rattles of this animal are, and yet how firmly they are united together, so as never to be

separated but by breaking them to pieces. One of those rattles singly, is incapable of producing sound, but the ringing of thirteen together, is sufficient to alarm the boldest man living." No wonder, then, that the rattlesnake came to symbolize American ideals.

DECEMBER 28

Listing and Delisting

On December 28, 1973, President Richard Nixon signed the Endangered Species Act (ESA) into law. The goal of the ESA is to protect species currently in danger of extinction ("Endangered") and those likely to become endangered within the foreseeable future ("Threatened") throughout all or a significant portion of their range. To do so, the ESA regulates activities that impact these species and their habitats. As of August 2017, 2392 species (1447 animals; 945 plants) were listed as endangered or threatened under the ESA.

Once a species is listed, recovery plans are developed and implemented. The status of each listed species is reviewed every five years. A species either remains on the list, or is removed if the threats to the species' survival have been eliminated or controlled and the species can survive on its own in the wild.

The West Virginia northern flying squirrel illustrates how the system works. In 1985, this subspecies was listed as Endangered because logging had destroyed much of its spruce-northern hardwood forest habitat. As the forests regenerated, populations increased; by 2008 the subspecies was delisted. In 2011, a lawsuit challenging the delisting was filed, and the subspecies was re-listed. Further study indicated that threats to the species' survival had been eliminated, and in 2013 a ruling reinstated the removal of the West Virginia northern flying squirrel from ESA protection. Flexibility and five-year reviews are keys to making the ESA the strongest federal law protecting US endangered and threatened plants and wildlife.

DECEMBER 29

Dinosaur's Cry

> What you hear is a voice. At first, it is the voice of the cello itself: the sound of wood carved, glued, polished, strung, and tuned in ways so as to replicate the tubes and chambers of a creature's innards. It is an animal sound, all furred and tendoned. In the slow passages it is elephantine, or older— a dinosaur's cry. In the fast ones it is equine, a steeple-chase run in stop-motion. But before long, the sound of this music, played this way, is a human sound.
>
> —PAUL ELIE, *Reinventing Bach*

Johann Sebastian Bach advocated a singing style when playing his music. Writer Paul Elie suggests that no one has done this better than Pablo Casals with his playing of Bach's Six Suites for Unaccompanied Cello. Casals, one of the world's greatest cellists, was born on December 29, 1876, in Catalonia, Spain.

If you like cello music, celebrate Casals' birthday today by listening to his renderings of Bach's Suites. See if you can hear what Elie describes—a transition as the suites progress from the voice of the wooden instrument itself, to elephantine and dinosaurian tones in the slow passages and equine tones in the fast ones, and finally to human song. The cello has an amazing capacity to mimic animal sounds, from mellow whale songs and somnolent cow moos, to the irritating barking and yipping of dogs, the loud, deep roar of a lion, and the human voice.

Part of Casals' greatness may have stemmed from the inspiration he found in nature. Casals began each day outside with a walk, followed by playing Johann Sebastian Bach. He claimed that walks outside filled him with awareness of the wonder of life.

DECEMBER 30

A Museum of Natural History

Nature-lovers of New York City found it inexcusable that by the 1860s a metropolis with the size, heterogeneity, and wealth of their city did not have a museum of natural history. Nor did any high school, college, or university in the city provide an education in nature study. In 1866, Albert S. Bickmore, a one-time student of Harvard zoologist Louis Agassiz, urged that a museum of natural history be established in New York City. His enthusiasm and energy catalyzed a group of New Yorkers to propose a plan.

On December 30, 1868, Andrew H. Green, comptroller of Central Park, received a letter signed by 19 persons, including businessman and philanthropist Theodore Roosevelt Sr.:

> Dear Sirs:
> A number of gentlemen having long desired that a great
> Museum of Natural History should be established in Central
> Park, and having now the opportunity of securing a rare
> and very valuable collection as a nucleus of such Museum,
> the undersigned wish to enquire if you are disposed to provide
> for its reception and development.

The following April a bill was signed to establish the American Museum of Natural History. The museum and its first collections, housed in the Arsenal building in Central Park, opened for viewing in 1871. Today the American Museum of Natural History is located across the street from Central Park. It occupies more than 2 million square feet (190,000 sq m), houses more than 32 million specimens and artifacts, and receives about 5 million visitors each year. The museum's mission statement reads: 'To discover, interpret, and disseminate—through scientific research and education—knowledge about human cultures, the natural world, and the universe." I think nineteenth-century nature-lovers of New York City would have approved.

DECEMBER 31

New Beginnings

The last day of the year is magical, bringing the promise of new beginnings. People around the world celebrate New Year's Eve with music, dancing, eating, and drinking, but beyond that, cultures differ in how they say farewell to the old year and welcome in the new. In Spain and many Central and South American countries, revelers eat 12 grapes, one with each strike of the chimes at midnight, to bring 12 months of prosperity. A Mexican custom is to write a list of all the unhappy or unfortunate events that happened in the past year, and burn the list at midnight. An Egyptian tradition involves smashing glass bottles and other breakables to signify the end of the year. In the Philippines, people create centerpieces of candies and round fruits. The circular shape of the fruits is believed to attract money during the coming year, and the candies represent hope for a sweeter year ahead.

New Year's Eve is a traditional time to make resolutions for the coming year. Many of us will vow to lose weight; become more understanding; get more exercise; or enjoy life more. What if we all

resolved to reaffirm our respect for life; appreciate the beauty and wonder of landscapes and the forces of nature; spend more time outside; and share our love of nature with others? Perhaps the other resolutions would come naturally. What would you see tomorrow with new eyes, something that you've never really appreciated before? Footprints in the snow, telling a story of predator stalking prey? The perfect six-fold radial symmetry of a snowflake on your mitten? Happy New Year!

Acknowledgments

I thank Christie Henry for believing in this project from its inception, and for her support and guiding hand with this and my previous four books with the University of Chicago Press, over nearly two decades. Christie, it has been a joy working with you. Thanks also to Miranda Martin and Mary Corrado at the University of Chicago Press for overseeing this book to completion.

Butch and Judy Brodie, Karen McKree, and Al Savitzky offered suggestions for themes and topics. Karen, my daughter and "unofficial editor," read early drafts of the entries and expertly guided the focus and organization of many. Karen, Al, and Anne Stark read the entire manuscript and offered valuable feedback. In the end, I alone am responsible for any errors or misinterpretations that might have crept into the entries. Bronwyn McIvor, artist extraordinaire, has given life to many of the plants and animals featured here. My husband, Al Savitzky, shared my enthusiasm, provided encouragement, and understood my sometimes obsessive focus on the project. I thank you all for your feedback, friendship, and support.

Finally, thanks to the dedicated professionals and laypersons working to protect our environment and the world's plants and animals; to the parents and teachers who encourage children to bond with nature; to the visual artists, musicians, and writers who share their passion for nature; to the scientists who help us to understand nature; and to Pachamama, Incan Earth Mother goddess, who, as life itself, protects and sustains us all.

Bibliography

January 1   Marshall, J. V. *Stories from the Billabong*. London: Frances Lincoln Children's Books, 2010.

January 2   Helm, R. R. "How Horseshoe Crabs May Have Saved Your Life." Deep Sea News, posted August 22, 2013. Accessed January 3, 2016. http://www.deepseanews.com/2013/08/how-horseshoe-crabs-may-have-saved-your-life/

January 3   Garter Snake. Massachusetts State Reptile. statesymbolsusa.org

January 4   Crump, M. *Eye of Newt and Toe of Frog, Adder's Fork and Lizard's Leg: The Lore and Mythology of Amphibians and Reptiles*. Chicago: University of Chicago Press, 2015.

January 5   Cox, D. T. C., D. F. Shanahan, H. L. Hudson, K. E. Plummer, G. M. Siriwardena, R. A. Fuller, K. Anderson, S. Hancock, and K. J. Gaston. "Doses of Neighborhood Nature: The Benefits for Mental Health of Living with Nature." *BioScience* 67 (2017): 147–155.

Leeming, D., and M. Leeming. *A Dictionary of Creation Myths*. New York: Oxford University Press, 1994.

January 6   Genetic Alliance, District of Columbia Department of Health. *Understanding Genetics: A District of Columbia Guide for Patients and Health Professionals*. "Classic Mendelian

Genetics (Patterns of Inheritance)." Accessed January 12, 2017. https://www.ncbi.nlm.nih.gov/books/NBK132145/

January 7   Thoreau, H. D. *Walden; or, Life in the Woods.* Boston: Ticknor and Fields, 1854.

January 8   Wallace, A. R. *A Narrative of Travels on the Amazon and Rio Negro.* New York: Dover, 1972. Originally published 1889.

January 9   Women & The Sea, The Mariners' Museum. "Mermaids." Accessed January 11, 2017. https://www.marinersmuseum.org/sites/micro/women/myths/mermaids.htm

January 10 Accomodation Direct. "Voodoo Day." Accessed January 10, 2017. http://www.benin-direct.com/activity/voodoo-day

Crump, M. *Eye of Newt and Toe of Frog, Adder's Fork and Lizard's Leg: The Lore and Mythology of Amphibians and Reptiles.* Chicago: University of Chicago Press, 2015.

January 11 The Aldo Leopold Foundation. http://www.aldoleopold.org

Leopold, A. *A Sand County Almanac; And Sketches Here and There.* New York: Oxford University Press, 1949.

Leopold, L. B., editor. *Round River: From the Journals of Aldo Leopold.* New York: Oxford University Press, 1993.

January 12 Beck, D. D. *Biology of Gila Monsters and Beaded Lizards.* Berkeley: University of California Press, 2005.

January 13 Martel, A., A. Spitzen-van der Sluijs, M. Blooi, et al. "*Batrachochytrium salamandrivorans* sp. nov, Causes Lethal Chytridiomycosis in Amphibians." *Proceedings of the National Academy of Sciences USA* 110 (2013): 15325–15329.

Zimmer, C. "U.S. Restricts Movement of Salamanders, for Their Own Good." *New York Times*, January 12, 2016.

January 14 Museum of Hoaxes. "The Surgeon's Photo." Accessed May 13, 2016. http://hoaxes.org/photo_database/image/the_surgeons_photo/

Scott, P., and R. Rines. "Naming the Loch Ness Monster." *Nature* 258, no. 5535 (1975): 466–468.

January 15 Nix, E. "Election 101: How Did the Republican and Democratic Parties Get Their Animal Symbols?" Ask History, July 7, 2015. Accessed May 13, 2016. http://www.history.com/news/ask-history/how-did-the-republican-and-democratic-parties-get-their-animal-symbols

January 16 The Dian Fossey Gorilla Fund International. "Dian Fossey—Biography." Accessed January 9, 2017. https://gorillafund.org/dian_fossey_bio

Mowat, F. *Woman in the Mists: The Story of Dian Fossey and the Mountain Gorillas of Africa.* New York: Warner, 1987.

January 17   Australian Koala Foundation. https://www.savethekoala.com

              Reed, A. W. *Aboriginal Legends: Animal Tales*. Victoria, Australia: Reed, 1978.

January 18   History.com Staff. "Cook Discovers Hawaii." Accessed May 13, 2016. http://www.history.com/this-day-in-history/cook-discovers-hawaii

January 19   Ives, M. "A Revered Turtle's Death Sets Hands Wringing." *New York Times*, January 23, 2016.

              Talk Vietnam, January 16, 2017. "Corpse of Hanoi's Legendary Turtle Preserved through Plastination." Accessed January 16, 2017. https://m.talkvietnam.com/2017/01/corpse-of-hanois-legendary-turtle-preserved-through-plastination/

January 20   ProFlowers, August 9, 2011. "Floriography: The Language of Flowers in the Victorian Era." Accessed June 14, 2016. http://www.proflowers.com/blog/floriography-language-flowers-victorian-era

January 21   Newman, M. "Baffling the Bandits." National Wildlife Federation, October 1, 2008. Accessed May 13, 2016. http://www.nwf.org/News-and-Magazines/National-Wildlife/Animals/Archives/2008/Science-Sleuths-How-Squirrels-Hide-Nuts-aspx

January 22   Poets.org. "Childe Harold's Pilgrimage [There is a pleasure in the pathless woods]," by George Gordon Byron. Accessed January 11, 2017. https://www.poets.org/poetsorg/poem/childe-harolds-pilgrimage-there-pleasure-pathless-woods

January 23   Woerner, A. "Is Cricket Flour the New Protein Powder?" *Daily Beast*, November 21, 2014. Accessed May 14, 2016. http://www.thedailybeast.com/articles/2014/11/21/is-cricket-flower-the-new-protein-powder.html

January 24   Crump, M. *Eye of Newt and Toe of Frog, Adder's Fork and Lizard's Leg: The Lore and Mythology of Amphibians and Reptiles*. Chicago: University of Chicago Press, 2015.

January 25   Tucker, A. "What Can Rodents Tell Us about Why Humans Love?" *Smithsonian Magazine*, February 2014. Accessed May 14, 2016. http://www.smithsonianmag.com/science-nature/what-can-rodents-tell-us-about-why-humans-love-180949441/?no-ist

January 26   Great Seal. "The Eagle, Ben Franklin, and the Wild Turkey." Accessed December 18, 2015. http://greatseal.com/symbols/turkey.html

January 27   The National Geographic Society. http://nationalgeographic.org

January 28   Dewey, J. O. *Rattlesnake Dance: True Tales, Mysteries, and Rattlesnake Ceremonies*. Honesdale, PA: Boyds Mills, 1997.

Klauber, L. M. *Rattlesnakes: Their Habits, Life Histories, and Influence on Mankind*. 2nd ed. Vol. 1 and 2. Berkeley: University of California Press, 1972.

"Timber Rattlesnakes Indirectly Benefit Human Health: Not-So-Horrid Top Predator Helps Check Lyme Disease." *ScienceDaily*, August 6, 2013. Accessed July 28, 2017. www.sciencedaily.com/releases/2013/08/130806091815.htm

January 29   America Comes Alive! "The First Seeing Eye Dog Is Used in America in 1928." Accessed May 13, 2016. http://america-comesalive.com/2012/06/25/how-a-dog-breeder

January 31   Space Answers. "Heroes of Space: Ham the Chimpanzee." Accessed May 13, 2016. http://www.spaceanswers.com/space-exploration/heroes-of-space-ham-the-chimpanzee/

February 1   Birds Choice. "February Is National Bird Feeding Month." Accessed January 10, 2017. http://www.birdschoice.com/backyard-birding/february-national-bird-feeding-month

February 2   Neatorama, February 2, 2011. "Groundhog Day or Hedgehog Day?" Accessed May 14, 2016. http://www.neatorama.com/2011/02/02/groundhog-day-or-hedgehog-day/

February 3   Muir, J. *The Yosemite*. New York: Century, 1912.

Yosemite National Park. National Park Service. "John Muir." Accessed March 10, 2017. https://www.nps.gov/yose/learn/historyculture/muir.htm

February 4   Stamets, P. *Mycelium Running: How Mushrooms Can Help Save the World*. Berkeley, CA: Ten Speed, 2005.

February 5   *A Piece of European Treasure*. "The Legend of Violet—the Flower." Accessed January 12, 2017. http://www.comenius-legends.blogspot.com/2010/07/legend-of-violet.html

February 7   Animal Planet. "Puppy Bowl." Accessed January 11, 2017. http://www.animalplanet.com/tv-shows/puppy-bowl/

February 8   Bates, H. W. *The Naturalist on the River Amazons*. 2 vol. London: John Murray, 1863.

February 9   Travel China Guide. "Chinese Zodiac." Accessed September 10, 2016. https://www.travelchinaguide.com/intro/social_customs/zodiac/

February 10   Shpansky, A. V., V. N. Aliyassova, and S. A. Ilyina. "The Quaternary Mammals from Kozhamzhar Locality (Pavlodar Region, Kazakhstan)." *American Journal of Applied Sciences* 13 (2016): 189–199.

February 11   Benchley, P. *Jaws*. New York: Doubleday, 1974.

February 12 Darwin, C. *The Autobiography of Charles Darwin: 1809–1882.* Edited by Nora Barlow. Rev. ed. New York: W. W. Norton, 1993. Originally published 1887.

February 13 WebExhibits. "Yaks, Butter, & Lamps in Tibet." Accessed June 18, 2016. http://www.webexhibits.org/butter/countries-tibet.html

February 14 Thelemapedia. "Aphrodite." Accessed January 13, 2017. http://www.thelemapedia.org/index.php/Aphrodite

February 15 Chawla, L. "Significant Life Experiences Revisited: A Review of Research on Sources of Environmental Sensitivity." *Environmental Education Research* 4 (1998): 369–382.

Sobel, D. *Beyond Ecophobia: Reclaiming the Heart in Nature Education.* Great Barrington, MA: Orion Society and the Myrin Institute, 1996.

February 16 Scoville, H. "Hugo de Vries. Early Life and Education." ThoughtCo. Updated November 29, 2015. Accessed February 20, 2017. http://evolution.about.com/od/scientists/p/Hugo-De-Vries.htm.

February 18 The Great Backyard Bird Count. eBird, powered by the Cornell Lab of Ornithology and the National Audubon Society. "About the GBBC." Accessed August 26, 2016. http://gbbc.birdcount.org/about/

February 21 British Trust for Ornithology. "National Nest Box Week." Accessed September 3, 2016. http://www.bto.org/about-birds/nnbw

Cornell Lab of Ornithology. "Learn about Nest Boxes and Nest Structures." Accessed September 3, 2016. http://nestwatch.org/learn/all-about-birdhouses/

February 22 CoinSite. "1913–1938 Buffalo Nickel." Accessed January 9, 2017. http://coinsite.com/1913–38-nickel-five-cents-buffalo/

February 23 Martin, L. C. *Wildlife Folklore.* Old Saybrook, CT: Globe Pequot, 1994.

"Rabbit Dance." *Indian Time*, February 19, 2015. Accessed May 24, 2016. http://www.indiantime.net/story/2015/02/19/cultural-corner/rabbit-dance/16906.html

February 24 PlantExplorers.com. "Joseph Banks 1743–1820." Accessed January 13, 2017. https://www.plantexplorers.com/explorers/biographies/banks/joseph-banks-01.htm

February 25 Stone, A. "'Shark Lady' Eugenie Clark, Famed Marine Biologist, Has Died." *National Geographic News*, February 25, 2015. Accessed July 26, 2016. http://news.nationalgeographic.com/2015/02/150225-eugenie-clark-shark-lady-marine-biologist-obituary-science/

February 26   EasyPetMD. "Newfoundland Dog 'Boatswain' Saves the Life of Napoleon Bonaparte." From Chambers's journal, Vol. 11–12, by William Chambers and Robert Chambers. Vol. 11, January–June 1849. Accessed January 13, 2017. http://www.easypetmd.com/newfoundland-dog-boatswain-saves-life-napoleon-bonaparte

Mott, M. "Guard Dogs: Newfoundland's Lifesaving Past, Present." *National Geographic News*, February 7, 2003. Accessed March 9, 2017. http://news.nationalgeographic.com/news/2003/02/0207_030207_newfies.html

February 27   Dybas, C. L. "Life after Ice: Polar Bears Struggle to Adapt to the New Normal." *Defenders of Wildlife* (Winter 2015). Accessed March 16, 2016. http://www.defenders.org/magazine/winter-2015/life-after-ice

February 28   Watson, J. D. *The Double Helix: A Personal Account of the Discovery of the Structure of DNA*. New York: Atheneum, 1968.

February 29   Sims, M. *Darwin's Orchestra: An Almanac of Nature in History and the Arts*. New York: Henry Holt, 1997.

March 1   Revolvy. "Ceres (Roman Mythology)." Accessed January 9, 2017. https://www.revolvy.com/main/index.php?s=Ceres%20(Roman%20mythology)&item_type=topic

March 2   Freethy, R. *Owls: A Guide for Ornithologists*. Kent, England: Bishopgate, 1992.

March 3   CITES. "What Is CITES?" Accessed January 10, 2017. https://cites.org/eng/disc/what.php

United Nations. "World Wildlife Day." Accessed July 21, 2016. http://www.un.org/en/events/wildlifeday/background.shtml

March 4   Coitir, N. M. *Ireland's Birds: Myths, Legends and Folklore*. West Link Park, Cork, Ireland: Collins, 2015.

March 5   Barnes, B. "Without Elephants, Ringling Circus Goes On." *New York Times*, July 13, 2016.

Mettler, K. "After 145 Years, Ringling Bros. Circus Elephants Perform for the Last Time." *Washington Post*, May 2, 2016.

March 6   Crump, M. *Mysteries of the Komodo Dragon: The Biggest, Deadliest Lizard Gives Up Its Secrets*. Honesdale, PA: Boyds Mills, 2010.

March 7   Thurber, A. R., W. J. Jones, and K. Schnabel. "Dancing for Food in the Deep Sea: Bacterial Farming by a New Species of Yeti Crab." *PLoS One* 6 (2011): e26243.doi:10.1371/journal.pone.0026243.

March 8	Danelski, D. "Hundreds of Threatened Desert Tortoises Will Be Moved from Marine Corps Base." *Press-Enterprise*, March 2, 2017. Accessed March 3, 2017. http://www.pe.com/articles/management-26698-corps-signed.html
	Lewis, D. "The Marine Corps Plans to Airlift over 1000 Desert Tortoises." *Smithsonian*, Smart News. smithsonian.com, March 10, 2016. Accessed March 11, 2016. http://www.smithsonianmag.com/smart-news/marine-corps-airlifting-desert-tortoises-new-training-grounds-180958315/#PL21zPqVz4T6kbAi.03
March 9	Dana Point Festival of Whales. "Festival of Whales 2016—Chronological Event Listing." Accessed July 20, 2016. http://festivalofwhales.com/dana-point-festival-of-whales-event-listing/
March 10	Galapagos Geology on the Web. "A Brief History of the Galapagos: Discovery, Pirates, and Whalers." Accessed January 10, 2017. http://www.geo.cornell.edu/geology/GalapagosWWW/Discovery.html
March 11	Armstrong, E. A. *The Life and Lore of the Bird in Nature, Art, Myth, and Literature*. New York: Crown, 1975.
	Cornell Lab of Ornithology. "The Search for the Ivory-Billed Woodpecker." Accessed March 13, 2017. http://www.birds.cornell.edu/ivory/
March 12	de Rohan, A. "Why Dolphins Are Deep Thinkers." *Guardian*, July 2, 2003. Accessed July 30, 2016. https://www.theguardian.com/science/2003/jul/03/research.science
March 13	Plinius, S. C. (Pliny). *Naturalis Historia. The Natural History of Pliny*. Translated by J. Bostock and H. T. Riley. Vol. 3, book 11, chap. 81. London: George Bell & Sons, 1892.
March 14	Leeming, D., and M. Leeming. *A Dictionary of Creation Myths*. New York: Oxford University Press, 1994.
	Michalski, K., and S. Michalski. *Spider*. London: Reaktion, 2010.
March 15	FLAP Canada. Accessed March 11, 2016. http://www.flap.org/who-we-are.php
March 16	Schaller, G. B. *The Last Panda*. Chicago: University of Chicago Press, 1993.
	Wildt, D. E., A. Zhang, H. Zhang, D. L. Janssen, and S. Ellis, eds. *Giant Pandas: Biology, Veterinary Medicine and Management*. Cambridge: Cambridge University Press, 2006.
March 17	Crump, M. *Eye of Newt and Toe of Frog, Adder's Fork and Lizard's Leg: The Lore and Mythology of Amphibians and Reptiles*. Chicago: University of Chicago Press, 2015.

March 18	Institute of Khmer Traditional Textiles. ikttearth.org.
March 19	Mission San Juan Capistrano. "Swallows Legend." Accessed November 17, 2015. https://www.missionsjc.com/about/swallows-legend/
March 20	Crump, M. L. "Anuran Reproductive Modes: Evolving Perspectives." *Journal of Herpetology* 49 (2015): 1–16.
	Crump, M. L. "The Many Ways to Beget a Frog." *Natural History* 86 (1977): 38–45.
March 21	Animal Tourism News. "Fitzgerald Wild Chicken Festival." Accessed December 23, 2015. http://animaltourism.com/news/2011/04/18/chicken-festival
	Autry, T. "The Cruelty of Rattlesnake Roundups." *Reptile Magazine*. Accessed January 17, 2016. http://www.reptilesmagazine.com/Venomous-Snakes/Rattlesnake-Roundups/
March 22	Orkneyjar. "The Selkie-Folk." Accessed January 10, 2017. http://www.orkneyjar.com/folklore/selkiefolk/
March 23	Porter, E. *"In Wildness Is the Preservation of the World."* San Francisco: Sierra Club, 1962.
March 24	Catesby Commemorative Trust. catesbytrust.org
	McBurney, H. *Mark Catesby's Natural History of America: The Watercolors from the Royal Library Windsor Castle.* London: Merrell Holberton, 1997.
March 25	We All Nepali. "Holi in Nepal." Accessed May 31, 2016. http://www.weallnepali.com/nepali-festivals/holi
March 26	Scott, M. "Conrad Gesner." Strangescience.net. Accessed July 29, 2016. http://www.strangescience.net/gesner.htm
March 27	Jonaitis, A. "Tlingit Halibut Hooks: An Analysis of the Visual Symbols of a Rite of Passage." *Anthropological Papers of the American Museum of Natural History* 57, part 1 (1981): 1–48.
March 28	Coitir, N. M. *Ireland's Mammals: Myths, Legends and Folklore.* West Link Park, Cork, Ireland: Collins, 2010.
March 29	Poetry Foundation. "I Wandered Lovely as a Cloud," by William Wordsworth. Accessed January 11, 2017. https://www.poetryfoundation.org/poems-and-poets/poems/detail/45521
March 30	Wigington, P. "From Egg-Laying Bunnies to Mad March Hares." About Religion. Updated March 11, 2016. Accessed June 16, 2016. http://paganwiccan.about.com/od/ostaramagic/a/RabbitFolklore.htm
March 31	Wise, B. "Earth Day: How Mother Nature Inspired Four Major Composers." WQXR, New York, April 21, 2015. Accessed

July 23, 2016. http://www.wqxr.org/story/earth-day-how-nature-inspired-major-composers/

April 1   Machiavelli, N., and D. Wootton. *The Prince*. Indianapolis: Hackett, 1995.

April 2   Duane, D. *So You Want to Be a Wizzard*. New York: Delacorte, 1983.

International Board on Books for Young People (IBBY). "International Children's Book Day." Accessed August 21, 2016. http://www.ibby.org/index.php?id=269

April 3   Ilgunas, K. "This Is Our Country. Let's Walk it." *New York Times*, April 24, 2016.

April 4   Guynup, S., and N. Ruggia. "Rats Rule at Indian Temple." *National Geographic News*, National Geographic Channel, June 29, 2004. Accessed September 8, 2016. http://news.nationalgeographic.com/news/2004/06/0628_040628_tvrats.html

April 5   National Park Service. "History of the Cherry Trees." Accessed February 22, 2017. https://www.nps.gov/subjects/cherry-blossom/history-of-the-cherry-trees.htm

April 6   Martin, L. C. *Wildlife Folklore*. Old Saybrook, CT: Globe Pequot, 1994.

McCracken, G. F. "Bats and the Netherworld." *Bats Magazine* 2 (1993). Accessed June 10, 2017. www.batcon/org/resources/media-education/bats-magazine/bat_article/589

April 7   Poetry Foundation. "The World Is Too Much with Us," by William Wordsworth. Accessed January 7, 2016. https://www.poetryfoundation.org/poems-and-poets/poems/detail/45564

Wordsworth, W., and S. T. Coleridge. *Lyrical Ballads with a Few Other Poems*. London: J. & A. Arch, 1798.

April 8   DABDAY.com. "Draw a Bird Day—April 8th." Accessed May 7, 2016. http://www.dabday.com/

Reed, A. W. *Aboriginal Myths: Tales of the Dreamtime*. New South Wales, Australia: Reed, 1978.

April 9   Florida Backroads Travel. "Sopchoppy Worm Gruntin Festival." Accessed February 3, 2016. http://www.florida-backroads-travel.com/sopchoppy-worm-gruntin-festival.html

April 10   Arbor Day Foundation. https://www.arborday.org

Tagore, R. *Fireflies: A Collection of Proverbs, Aphorisms and Maxims*. Wakefield, RI: Asphodel, 2007.

April 11   John-Keats.com. "I Stood Tip-toe upon a Little Hill," by John
                Keats. Accessed January 13, 2017. http://www.john-keats.
                com/gedichte/i_stood_tip-toe.htm

April 12   Deming, A. H. *The Monarchs: A Poem Sequence*. Baton Rouge:
                Louisiana State University Press, 1997.

April 13   Smith, J. "Spotlight: Alaska Hummingbird Festival in Ket-
                chikan, Alaska." Recreation.gov. Accessed September 27,
                2017. https://www.recreation.gov/marketing.do?goto=acm/
                Explore_And_More/exploreArticles/Spotlight_Alaska_
                Hummingbird_Festival_in_Ketchikan_Alaska.htm

April 14   Wilkinson, G. S. "Food Sharing in Vampire Bats." *Scientific
                American* (1990): 76–82.

April 15   Seawright, C. "Bast, Perfumed Protector, Cat Goddess." Tour
                Egypt. Accessed May 27, 2016. http://www.touregypt.net/
                godsofegypt/bast2.htm

April 16   Williams, J. A., C. Podeschi, N. Palmer, P. Schwadel, and D.
                Meyler. "The Human-Environment Dialog in Award-
                Winning Children's Picture Books." *Sociological Inquiry* 82
                (2012): 145–159.

April 17   Bayou Teche Black Bear Festival. Accessed July 31, 2017. http://
                www.bayoutechebearfest.org

                Bieder, R. E. *Bear*. London: Reaktion, 2005.

April 18   del Giudice, V. "First Lady Pitches 'Pennies for Pandas.'" UPI
                (United Press International), March 26, 1984. Accessed
                August 31, 2016. http://www.upi.com/Archives/1984/03/26/
                First-lady-pitches-Pennies-for-Pandas/3910449125200/

April 19   Tamura, A. "California Condor Feathers Tell Harrowing Tale of
                Struggle and Survival." June 16, 2014. Accessed January 11,
                2017. http://blog.condorwatch.org/2014/06/16/california-
                condor-feathers-tell-harrowing-tale-of-struggle-and-
                survival-guest-post-by-alex-tamura/

                US Fish & Wildlife Service, Pacific Southwest Region. Califor-
                nia Condor Recovery Program. "California Condor." Last
                updated September 28, 2016. Accessed January 9, 2017.
                http://www.fws.gov/cno/es/CalCondor/Condor.cfm

April 20   Clark, P. "No Fish Story: Sandwich Saved His McDonald's."
                Cincinnati Enquirer, February 20, 2007. USA Today. Ac-
                cessed January 15, 2016. http://usatoday30.usatoday.com/
                money/industries/food/2007-02-20-fish2-usat_x.htm

April 21   Garcia, B. "Romulus and Remus." Ancient History Encyclope-
                dia, October 4, 2013. Accessed May 31, 2016. http://www.
                ancient.eu/Romulus_and_Remus/

April 22   Turnage, W. "Ansel Adams, Photographer." The Ansel Adams Gallery. Accessed March 12, 2017. http://anseladams.com/about-ansel-adams/ansel-adams-biography/

April 23   Freeman, E. H. "Children Thrive in Outdoor Preschool." *Herald Journal*, Writers on the Range (High Country News), January 10, 2016.

Louv, R. *Last Child in the Woods: Saving Our Children from Nature-Deficit Disorder*. New York: Workman, 2005.

Malone, K., and S. Waite. "Student Outcomes and Natural Schooling." Plymouth: Plymouth University. Accessed December 17, 2016. Available online: http://www.plymouth.ac.uk/research/oelres-net

April 24   Religious Holidays & Festivals. "Mahavir Jayanti: Jains Mimic Tirthankar with Nonviolence." April 25, 2013. Accessed January 2, 2016. http://www.readthespirit.com/religious-holidays-festivals/tag/jain/

April 25   Frison, M. "Celebrate World Penguin Day!" Ian Somerhalder Foundation. Accessed May 24, 2016. http://www.isfoundation.com/news/celebrate-world-penguin-day

McSweeney, M. "Tawaki—The Rainforest Penguin." 100% Pure New Zealand. Accessed May 24, 2016. http://www.newzealand.com/us/article/tawaki-the-rainforest-penguin/

April 26   Encyclopedia of World Biography. "John James Audubon Biography." Accessed August 29, 2016. http://www.notablebiographies.com/An-Ba/Audubon-John-James.html

April 27   Collins, J. P., and M. L. Crump. *Extinction in Our Times: Global Amphibian Decline*. New York: Oxford University Press, 2009.

Crump, M. *Eye of Newt and Toe of Frog, Adder's Fork and Lizard's Leg: The Lore and Mythology of Amphibians and Reptiles*. Chicago: University of Chicago Press, 2015.

Save the Frogs. "Save the Frogs Day." Accessed September 6, 2016. http://savethefrogs.com/d/day/index.html

April 28   Furman, B. L. "The Development of Byetta (Exenatide) from the Venom of the Gila Monster as an Anti-diabetic Agent." *Toxicon* 59 (2012): 464–471.

April 29   Center for Disease Control and Prevention. "Tetrodotoxin Poisoning Associated with Eating Puffer Fish Transported from Japan–California, 1996." *Morbidity and Mortality Weekly Report* 45 (1996): 389–391. Accessed August 29, 2016. http://www.cdc.gov/mmwr/preview/mmwrhtml/00041514.htm

Myers, C. W., J. W. Daly, and B. Malkin. "A Dangerously Toxic New Frog (*Phyllobates*) Used by Emberá Indians of Western Colombia, with Discussion of Blowgun Fabrication and Dart Poisoning." *Bulletin of the American Museum of Natural History* 161 (1978): 309–365.

April 30   Pikulicka-Wilczewska, A. "A Day at the Races in Horse-Mad Turkmenistan." *Guardian*, November 22, 2015. Accessed May 31, 2016. http://www.theguardian.com/world/2015/nov/22/turkmenistan-horses-akhal-teke-breed-races

May 1   Wigington, P. "Floralia: The Roman May Day Celebration." About Religion. Updated April 26, 2016. Accessed September 7, 2016. http://paganwiccan.about.com/od/beltanemayday/p/Floralia.htm

May 2   Habeeb.com. "On Clothes," from *The Prophet* by Khalil Gibran. Accessed January 13, 2017. http://www.katsandogz.com/onclothes.html

May 3   History.com Staff. "War Animals from Horses to Glowworms: 7 Incredible Facts." History.com. Accessed September 7, 2016. http://www.history.com/news/war-animals-from-horses-to-glowworms-7-incredible-facts

May 4   United Poultry Concerns. "International Respect for Chickens Day Celebrates Compassion for Chickens." Accessed August 21, 2016. http://www.upc-online.org/nr/160428ircd.html

May 5   "Tokyo Hedgehog Café Brings More Kicks than Pricks." *Irish Times*, April 7, 2016. Accessed April 8, 2016. http://www.irishtimes.com/news/world/asia-pacific/tokyo-hedgehog-café-brings-more-kicks-than-pricks-1.2601891

May 6   World of Flowering Plants. "Legends and Facts about the Lily of the Valley." November 12, 2014. Accessed June 23, 2016. http://worldoffloweringplants.com/legends-facts-lily-valley/

May 7   Cobb, W. D. "The Pearl of Allah." *Natural History* 44 (1939): 197–202.

May 8   Carr, A. *The Windward Road: Adventures of a Naturalist on Remote Caribbean Shores.* Tallahassee: University Press of Florida, 1979. Originally published 1955.

Davis, F. R. *The Man Who Saved Sea Turtles: Archie Carr and the Origins of Conservation Biology.* New York: Oxford University Press, 2007.

Ewel, J. J. "Awards. Eminent Ecologist." *Bulletin of the Ecological Society of America* 69 (1988): 24–25.

May 9	Digest of Federal Resource Laws of Interest to the US Fish and Wildlife Service. "Migratory Bird Treaty Act of 1918." Accessed December 31, 2016. https://www.fws.gov/laws/lawsdigest/MIGTREA.HTML
	Runge, C. A., J. E. M. Watson, S. H. M. Butchart, J. O. Hanson, H. P. Possingham, and R. A. Fuller. "Protected Areas and Global Conservation of Migratory Birds." *Science* 350 (2015): 1255–1258.
May 10	Grahame, K. *The Wind in the Willows*. New York: Heritage Illustrated Bookshelf, 1940. Originally published 1908.
	What's On? Bodleian Libraries. University of Oxford. "The Original Wind in the Willows." Accessed August 31, 2016. http://www.bodleian.ox.ac.uk/whatson/whats-on/online/witw/letters
May 11	Christie, A. *The Hound of Death and Other Stories*. London: Odhams, 1933.
	Crump, M. *Headless Males Make Great Lovers & Other Unusual Natural Histories*. Chicago: University of Chicago Press, 2005.
May 12	Biodiversity Heritage Library. Notes & News from the BHL Staff. "Albertus Seba's Cabinet of Wonder and Awe." January 17, 2013. Accessed January 11, 2017. http://blog.biodiversitylibrary.org/2013/01/curiositycabinet.html
	Seba, A. *Cabinet of Natural Curiosities*. [Based on the copy in the Koninklijke Bibliotheek, The Hague.] Cologne, Germany: Taschen, 2011.
May 14	Hawkins, V. "The U.S. Army's 'Camel Corps' Experiment." National Museum United States Army, July 16, 2014. Accessed January 9, 2017. https://armyhistory.org/the-u-s-armys-camel-corps-experiment/
May 15	Van Huis, A., M. Dicke, and J. A. A. van Loon. "Insects to Feed the World." *Journal of Insects as Food and Feed* 1 (2015): 3–5. Accessed September 27, 2017. doi:10.3920/JIFF2015.x002.
May 16	Khan, M. L. "Center for the Conservation of Biodiversity: Sacred Groves in India." XII World Forestry Congress, 2003. Accessed January 2, 2016. http://www.fao.org/docrep/ARTICLE/WFC/XII/0509-A1.HTM.
May 17	Metropolitan Museum of Art. "The Heart of the Andes." Accessed March 12, 2017. http://www.metmuseum.org/art/collection/search/10481.
	Wulf, A. *The Invention of Nature: Alexander von Humboldt's New World*. New York: Alfred A. Knopf, 2015.

May 18   Dale, V. H., C. M. Crisafulli, and F. J. Swanson. "25 Years of
            Ecological Change at Mount St. Helens." *Science* 308
            (2005): 961–962.

            Pacific Northwest Research Station, USDA Forest Service.
            "Mount St. Helens 30 Years Later: A Landscape Reconfig-
            ured." *PNW Science Update* 19 (2010): 1–11.

            Thompson, A. "Mount St. Helens Still Recovering 30 Years
            Later." Live Science, May 17, 2010. Accessed May 31,
            2016. http://www.livescience.com/6450-mount-st-helens-
            recovering-30-years.html

May 20   Columbia University School of the Arts. "Dürer's Rhinoceros."
            Accessed August 5, 2016. http://arts.columbia.edu/du-
            rersrhinoceros

May 21   Woods, J. "Meet Mary Reynolds, the Plant Whisperer Who
            Dared to Be Wild." *Telegraph*, September 4, 2016. Accessed
            November 15, 2016. http://www.telegraph.co.uk/garden-
            ing/chelsea-flower-show/meet-mary-reynolds-the-plant-
            whisperer-who-dared-to-be-wild/

May 22   Convention on Biological Diversity. "History of the Conven-
            tion." Accessed July 2, 2016. https://www.cbd.int/history/

            Convention on Biological Diversity. "International Day for
            Biological Diversity—22 May." Accessed January 10, 2017.
            https://www.cbd.int/idb/

May 23   Crump, M. *Eye of Newt and Toe of Frog, Adder's Fork and
            Lizard's Leg: The Lore and Mythology of Amphibians and
            Reptiles.* Chicago: University of Chicago Press, 2015.

May 24   Tourism of Cambodia. "Royal Ploughing Ceremony." Accessed
            September 27, 2017. www.tourismcambodia.com/tripplan-
            ner/events-in-cambodia/royal-ploughing.htm

May 25   US Fish & Wildlife Service International Affairs. "Lacey Act."
            Accessed August 5, 2016. https://www.fws.gov/internation-
            al/laws-treaties-agreements/us-conservation-laws/lacey-
            act.html

            Wisch, R. F. "Overview of the Lacey Act (16 U.S.C. SS 3371–
            3378)." Michigan State University College of Law, 2003.
            Animal Legal & Historical Center. Accessed August 8,
            2016. https://www.animallaw.info/article/overview-lacey-
            act-16-usc-ss-3371–3378

May 26   Cousteau, J. Y., with F. Dumas. *The Silent World: A Story of Un-
            dersea Discovery and Adventure.* New York: HarperCollins,
            1953.

May 27   West, M. J., and A. P. King. "Mozart's Starling." *American Sci-
            entist* 78 (1990): 106–114.

May 28   Rawlings, M. K. *The Yearling*. New York: Charles Scribner's
              Sons, 1938.

              Whooping Crane Eastern Partnership. www.bringbackthe-
              cranes.org

May 29   Buchen, L. "May 29, 1919: A Major Eclipse, Relatively Speak-
              ing." Wired, May 29, 2009. Accessed January 12, 2017.
              https://www.wired.com/2009/05/dayintech_0529/

May 30   Share, J. "Hopi Corn, Kachina Rain and Lessons from the
              Past," posted November 25, 2011. Accessed June 17, 2016.
              Blogspot.com. http://written-in-stone-seen-through-my-
              lens.blogspot.com/2011/11/hopi-corn-and-lessons-from-
              past.html

              Stoller, M. L. "Birds, Feathers, and Hopi Ceremonialism."
              Expedition Magazine 33 (1991): 35–45.

May 31   The International Ecotourism Society. www.ecotourism.org

June 1   Coitir, N. M. *Ireland's Birds: Myths, Legends and Folklore*. West
              Link Park, Cork, Ireland: Collins, 2015.

June 2   Kilgannon, C. "A Rite to Improve Karma for Man, Creature
              and, Now, the Environment." *New York Times*, November
              28, 2015.

              O'Brien, B. "Saga Dawa or Saka Dawa: Holy Month for Bud-
              dhists." ThoughtCo. Accessed January 6, 2017. http://
              buddhism.about.com/od/buddhistholidays/fl/Saga-Dawa-
              or-Saka-Dawa.htm

June 3   American Museum of Natural History. "James Hutton: The
              Founder of Modern Geology." Accessed December 5, 2016.
              http://www.amnh.org/explore/resource-collections/earth-
              inside-and-out/james-hutton-the-founder-of-modern-
              geology/

June 4   France.fr. "Roquefort, the King of Cheese." Accessed August 14,
              2016. http://us.france.fr/en/information/roquefort-king-
              cheese

June 5   United Nations. "World Environment Day." Accessed March
              13, 2017. http://www.un.org/en/events/environmentday/
              index.shtml

June 6   Poetry Foundation. "A Red, Red Rose," by Robert Burns.
              Accessed January 13, 2017. https://www.poetryfoundation.
              org/poems-and-poets/poems/detail/43812

June 7   Castro, G., and J. P. Myers. "Shorebird Predation of Eggs of
              Horseshoe Crabs during Spring Stopover on Delaware Bay."
              Auk 110 (1993): 927–930.

June 8   Prager, E. *Sex, Drugs, and Sea Slime: The Ocean's Oddest*

Creatures and Why They Matter. Chicago: University of Chicago Press, 2011.

June 9 Sadler, R. W. "Seagulls, Miracle of." In *Encyclopedia of Mormonism: The History, Scripture, Doctrine, and Procedure of the Church of Jesus Christ of Latter-day Saints*, edited by D. H. Ludlow, 1287–1288. New York: Macmillan, 1992. Accessed August 13, 2016. http://eom.byu.edu/index.php/Seagulls,_Miracle_of

June 10 Wilson, E. O. *Biophilia: The Human Bond with Other Species*. Cambridge, MA: Harvard University Press, 1984.

June 11 Geggel, L. "Is It Possible to Clone a Dinosaur?" LiveScience, April 28, 2016. Accessed January 29, 2017. http://www.livescience.com/54574-can-we-clone-dinosaurs.html

June 12 Seawright, C. "Mut, Mother Goddess of the New Kingdom, Wife of Amen, Vulture Goddess." Tour Egypt. Last updated June 11, 2011. Accessed May 31, 2016. http://www.tour-egypt.net/featurestories/mut.htm

June 13 Poetry Foundation. "The Wild Swans at Coole," by William Butler Yeats. Accessed January 7, 2017. https://www.poetry-foundation.org/poems-and-poets/poems/detail/43288

June 14 Brooks, C. "Dive In to National Rivers Month." American Forests, June 12, 2014. Accessed June 22, 2016. http://www.americanforests.org/blog/dive-in-to-national-rivers-month/

June 15 Kim, S. W. "Light Up the Night with a Firefly Festival." *Japan Times*, June 5, 2014. Accessed November 12, 2015. http://www.japantimes.co.jp/culture/2014/06/05/events/events-outisde-tokyo/light-up-the-night-with-a-firefly-festival/#.VkUYmYSDJVs

June 16 Carr, A. *So Excellent a Fishe: A Natural History of Sea Turtles*. Garden City, NY: Natural History Press, for American Museum of Natural History, 1967.

June 17 Lankester, R. "On *Okapia johnstoni*." *Proceedings of the Royal Society of London* 2 (1901): 279–281.

Lankester, R. "On *Okapia*, a New Genus of Giraffidae, from Central Africa." *Transactions of the Zoological Society of London* 16 (1902): 279–314.

June 18 Liberty Coin & Currency. "The Top Ten Most Famous Animal Coins." Posted on August 2, 2014. Accessed August 17, 2016. http://libertycoinandcurrency.com/blog/the-top-ten-most-famous-animal-coins/

June 19 Crump, M. *The Mystery of Darwin's Frog*. Honesdale, PA: Boyds Mills, 2013.

June 20   American Presidency Project. "Ronald Reagan: Proclamation 4893—Bicentennial Year of the American Bald Eagle and National Bald Eagle Day," January 28, 1982. Online by Gerhard Peters and John T. Woolley, The American Presidency Project. Accessed January 8, 2017. http://www.presidency.ucsb.edu/ws/?pid=42831

June 21   Antiquity Now. "The Summer Solstice: From Ancient Celebration to a Modern Day at the Beach." Accessed March 8, 2017. https://antiquitynow.org/2014/06/19/the-summer-solstice-from-ancient-celebration-to-a-modern-day-at-the-beach/

June 22   Popovic, M. "Nalukataq." traditionscustoms.com. Accessed May 11, 2016. http://www.traditionscustoms.com/lifestyle/nalukataq

June 23   MacCoitir, N. *Irish Wild Plants: Myths, Legends & Folklore.* West Link Park, Cork, Ireland: Collins, 2006.

June 24   Nicholls, H. "The Legacy of Lonesome George." *Nature* 487 (2012): 279–280. doi:10.1038/487279a

           Nicholls, H. "A Giant in New York. Lonesome George Returns from the Dead." *Guardian*, September 17, 2014. Accessed September 6, 2016. https://www.theguardian.com/science/animal-magic/2014/sep/17/lonesome-george-taxidermy-new-york

June 25   Civil War Talk. "National Catfish Day." Accessed March 19, 2017. http://civilwartalk.com/threads/national-catfish-day.125348/

June 27   Mythphile. "Science May Explain Why Egyptians Worshiped Dung Beetle as Sun God." Accessed September 7, 2016. http://www.mythphile.com/2012/01/ancient-egyptian-scarab-beetle/

June 28   Associations of Zoos and Aquariums. "FrogWatch USA." Accessed January 9, 2017. https://www.aza.org/frogwatch

June 29   Strawberry Festival. Accessed September 7, 2016. http://www.cedarburg.org/event/1465436-strawberry-festival-2016

June 30   Davis, E. "Horseback Shrimp Fishing Fades in Belgium." *New York Times*, August 31, 2007.

           Oostduinkerke.com. "Oostduinkerke, Beach of the Horse Fisherman." Accessed September 10, 2016. https://oostduinkerke.com/en/oostduinkerke.php

July 1   World Association of Zoos and Aquariums. http://waza.org

July 2   Nature's Calendar. http://www.naturescalendar.org

July 3   Galasso, S. "When the Last of the Great Auks Died, It Was by the Crush of a Fisherman's Boot." Smithsonian.com. Accessed December 8, 2016. http://www.smithsonianmag. com/smithsonian-institution/with-crush-fisherman-boot-the-last-great-auks-died-180951982/

July 4   Eisner, T., and D. J. Aneshansley. "Spray Aiming in Bombardier Beetles: Jet Deflection by the Coanda Effect." *Science* 215 (1982): 83–85.

July 5   AnimalResearch.Info. "Cloning Dolly the Sheep." Accessed December 15, 2016. http://www.animalresearch.info/en/ medical-advances/timeline/cloning-dolly-the-sheep/

July 6   Andersen, H. C. "The Storks." Translated by M. R. James. In: *Hans Christian Andersen's Forty-Two Stories.* London: Faber and Faber, 1930. "The Stork" originally published 1838.

Rowland, B. *Birds with Human Souls: A Guide to Bird Symbolism.* Knoxville: University of Tennessee Press, 1978.

July 7   Hart, H. "July 7, 1550: Europeans Discover Chocolate." Wired. Accessed July 19, 2016. http://www.wired. com/2010/07/0707chocolate-introduced-europe/

July 8   Florida Keys News. "Lower Keys Underwater Music Festival to Promote Reef Protection." Accessed January 8, 2017. http://www.fla-keys.com/news/news.cfm?sid=8931

July 9   ChewingCane.com. "What Is Sugarcane? History of Sugar Cane." Accessed September 4, 2016. http://www.chewing-cane.com/sugarcane_history.html

July 10   United States Department of Agriculture. "History of the BW-CAW." Accessed December 30, 2016. https://www.fs.usda. gov/detail/superior/specialplaces/?cid=stelprdb5127455.

July 11   Gudernatsch, J. F. "Feeding Experiments on Tadpoles. I. The Influence of Specific Organs Given as Food on Growth and Differentiation. A Contribution to the Knowledge of Organs with Internal Secretion." *Wilhelm Roux Arch. Entwicklungsmech. Organismen.* 35 (1912): 457–483.

July 12   Cow Appreciation Day. Accessed June 13, 2016. http://www. cute-calendar.com/event/cow-appreciation-day/19658. html

National Hindu Students' Forum (UK). "Why Do Hindus Worship the Cow?" Accessed June 13, 2016. http://www.nhsf. org.uk/2007/05/why-do-hindus-worship-the-cow/

July 13   Ancient Egypt Online. "Ra. The Sun God of Egypt." Accessed December 21, 2016. http://www.ancient-egypt-online.com/ egyptian-god-ra.html

July 14   Crawford, D. *Shark*. London: Reaktion, 2008.

            Peachin, M. L. *Underwater Encounters: What You Should Know
                about Sharks*. Amazon Digital Services: Peachin Adven-
                ture, 2014.

            Quammen, D. *Monster of God: The Man-Eating Predator in the
                Jungles of History and the Mind*. New York: W. W. Norton,
                2003.

July 15   Mishima, S. "Japan's Obon Festival: Everything You Need to
                Know." About Travel, updated August 18, 2016. Accessed
                January 8, 2017. http://gojapan.about.com/cs/japanesefes-
                tivals/a/obonfestival.htm

July 16   Crump, M. *Eye of Newt and Toe of Frog, Adder's Fork and
                Lizard's Leg: The Lore and Mythology of Amphibians and
                Reptiles*. Chicago: University of Chicago Press, 2015.

            Morgan, D. *Snakes in Myth, Magic, and History: The Story of a
                Human Obsession*. Westport, CT: Praeger, 2008.

July 17   Carroll, R. "Dung Loaming: How Llamas Aided the Inca Em-
                pire." *Guardian*, May 22, 2011. Accessed February 23, 2017.
                https://www.theguardian.com/world/2011/may/22/incas-
                llama-manure-crops

            Llamapaedia. "Llama Origin & Domestication." Accessed
                January 12, 2016. http://www.llamapaedia.com/origin/
                domestic.html

July 18   eBird. "Cornell Lab of Ornithology Young Birders Event 2017."
                Accessed September 27, 2017. http://ebird.org/content/
                ebird/news/yb2016/

July 19   Yarmouth Clam Festival. Accessed January 8, 2016. http://
                www.clamfestival.com/

July 21   Reef Awareness Week. Accessed September 9, 2016. http://
                www.fla-keys.com/news/news.cfm?sid=205

July 22   Bartram, W. *Travels through North and South Carolina,
                Georgia, East and West Florida, the Cherokee Country,
                the Extensive Territories of the Muscogulges, or Creek
                Confederacy, and the Country of the Chactaws; Containing
                an Account of the Soil and Natural Productions of Those
                Regions; Together with Observations on the Manners of the
                Indians*. Philadelphia: James & Johnson, 1791.

            Romantic Natural History: A Survey of Relationships between
                Literary Works and Natural History in the Century before
                Charles Darwin's *On the Origin of Species* (1859). "Wil-
                liam Bartram (1739–1823)." Accessed September 15, 2016.
                http://users.dickinson.edu/~nicholsa/Romnat/bartram.
                htm

July 23   welcomearmenia.com. "Armenia. Vardavar." Accessed September 27, 2017. www.welcomearmenia.com/armenia/vardavar.

July 24   Baldwin, L. "Golden Retrievers Go 'Home' for Gathering in Scottish Highlands." PBS NewsHour, The Rundown, August 7, 2013. Accessed May 2, 2016. http://www.pbs.org/newshour/rundown/golden-retriever-gathering/

July 25   Leach, M., ed. *Funk & Wagnalls Standard Dictionary of Folklore, Mythology and Legend*. New York: Funk & Wagnalls, 1972.

July 26   Crump, M. *Eye of Newt and Toe of Frog, Adder's Fork and Lizard's Leg: The Lore and Mythology of Amphibians and Reptiles*. Chicago: University of Chicago Press, 2015.

July 27   UnmissableJAPAN.com. "Cormorant Fishing." Accessed July 26, 2016. http://www.unmissablejapan.com/events/ukai

July 28   Plocek, K. "Paleontologist's Wandering Skull." mental_floss. Accessed July 24, 2016. http://mentalfloss.com/article/60125/edward-drinker-cope-and-story-paleontologists-wandering-skull

July 29   Gilroy Garlic Festival. Accessed August 28, 2016. http://gilroygarlicfestival.com/

July 30   Hoare, B. "Britain's National Species Revealed." *Discover Wildlife*, BBC Wildlife magazine, July 30, 2013. Accessed January 26, 2016. http://www.discoverwildlife.com/british-wildlife/britains-national-species-revealed

July 31   Erens, H., M. Boudin, F. Mees, B. B. Mujinya, G. Baert, M. Van Strydonck, P. Boeckx, and E. Van Ranst. "The Age of Large Termite Mounds—Radiocarbon Dating of *Macrotermes falciger* Mounds of the Miombo Woodland of Katanga, DR Congo." *Palaeogeography, Palaeoclimatology, Palaeoecology* 435 (2015): 265–271.

         Walker, M. "2000-Year-Old Termite Mound Found." BBC, July 31, 2015. Accessed August 13, 2016. http://www.bbc.com/earth/story/20150729–2000-year-old-termite-mound-found

August 1   Reynolds, J. "Save the Whales, Save Ourselves: Why Whales Matter." Natural Resources Defense Council, October 15, 2013. Accessed July 19, 2016. https://www.nrdc.org/experts/joel-reynolds/save-whales-save-ourselves-why-whales-matter

August 2   Carroll, L. *Alice's Adventures in Wonderland*. New York: D. Appleton, 1866. Originally published 1865.

August 3	Jackson, L. "Olympic Girl Power: The Incredible Story of Lis Hartel." *Horse Nation*, November 17, 2014. Accessed August 18, 2016. http://www.horsenation.com/2014/11/17/olympic-girl-power-the-incredible-story-of-lis-hartel/
	"Jubilee, a Post-war Dressage Hero." *Eurodressage*, October 22, 2010. Accessed February 24, 2017. http://www.euro-dressage.com/equestrian/2010/10/22/jubilee-post-war-dressage-hero
August 4	About International Assistance Dog Week. Accessed December 8, 2016. http://www.assistancedogweek.org/about/
August 5	Moss, L. "Elephant and Dog Are Best Friends." Mother Nature Network. September 18, 2013. Accessed February 26, 2017. http://www.mnn.com/family/pets/stories/elephant-and-dog-are-best-friends
August 6	Poetry Foundation. "The Eagle," by Alfred, Lord Tennyson. Accessed January 11, 2017. https://www.poetryfoundation.org/poems-and-poets/poems/detail/45322
	Reference.com. "How High Can an Eagle Fly?" Accessed February 26, 2017. https://www.reference.com/pets-animals/high-can-eagle-fly-98ca31be0fecef5d#
August 7	gaiatheory.org. "Gaia Theory: Model and Metaphor for the 21st Century." Accessed June 21, 2016. http://www.gaiatheory.org/overview/
August 8	Crump, M. *Eye of Newt and Toe of Frog, Adder's Fork and Lizard's Leg: The Lore and Mythology of Amphibians and Reptiles.* Chicago: University of Chicago Press, 2015.
August 9	Wildlife Spotter. Accessed September 11, 2016. https://wildlifespotter.net.au/about/
August 10	World Lion Day. Accessed January 11, 2017. https://worldlionday.com/the-campaign/
August 11	Coitir, N. M. *Ireland's Mammals: Myths, Legends and Folklore.* West Link Park, Cork, Ireland: Collins, 2010.
August 12	The Asian Elephant Art & Conservation Project. http://elephantart.com.
	World Elephant Day. www.worldelephantday.org
August 13	Nova Roma. "Nemoralia." Accessed January 2, 2016. http://www.novaroma.org/nr/Nemoralia
August 14	Crump, M. *Eye of Newt and Toe of Frog, Adder's Fork and Lizard's Leg: The Lore and Mythology of Amphibians and Reptiles.* Chicago: University of Chicago Press, 2015.
	Martin, J. *Masters of Disguise: A Natural History of Chameleons.* New York: Facts on File, 1992.

August 15   Beebe, W. *Half Mile Down*. New York: Duell Sloan Pearce, 1951.

Official William Beebe Website. "Bathysphere." Accessed May 6, 2016. https://sites.google.com/site/cwilliambeebe/Home/bathysphere

August 16   Crystalinks. "Annual Flooding of the Nile." Accessed August 19, 2016. http://www.crystalinks.com/floodingnile.html

August 17   Schaul, J, C. "Black Cat Appreciation Day: Do You Know Your Melanistic Cats?" Posted on Cat Watch, August 17, 2013. Accessed July 2, 2016. http://voices.nationalgeographic.com/2013/08/17/black-cat-appreciation-day-do-you-know-your-melanistic-cats/

August 18   Donlan, J., H. W. Greene, J. Berger, et al. "Re-wilding North America." *Nature* 436 (2005): 913–914.

Galetti, M. "Parks of the Pleistocene: Recreating the Cerrado and the Pantanal with Megafauna." *Natureza & Conservação* 2 (2004): 93–100.

August 20   Coitir, N. M. *Ireland's Mammals: Myths, Legends and Folklore*. West Link Park, Cork, Ireland: Collins, 2010.

August 21   Bromwich, J. E. "The Demons of Darkness Will Eat Men, and Other Solar Eclipse Myths." *New York Times*, August 19, 2017.

Fonseca, F. "Tribes Look for Renewal from Eclipse." *Herald Journal*, August 20, 2017.

August 22   Caserita.info. "Pachamama: Mother Earth." Accessed September 9, 2016. http://info.handicraft-bolivia.com/Pachamama-Mother-Earth-a346

August 23   Wigington, P. "What Was the Vulcanalia?" About Religion, updated August 31, 2016. Accessed September 11, 2016. http://paganwiccan.about.com/od/LammasFolklore/p/The-Vulcanalia.htm

August 24   Comstock, A. B. *Handbook of Nature Study*. 6th ed. Ithaca, NY: Comstock Publishing/Cornell University Press, 1986. Originally published 1911.

Cornell University Natural History Collections. "Anna Comstock." Accessed March 12, 2017. http://naturalhistorycollections.cornell.edu/insect/comstock.html

August 25   National Park Service. "History." Accessed January 10, 2017. https://www.nps.gov/aboutus/history.htm

National Park Service. "National Park System Timeline (Annotated)." Accessed January 10, 2017. https://www.nps.gov/parkhistory/hisnps/NPSHistory/timeline_annotated.htm

August 26   Insurance Information Institute. "Facts + Statistics: Pet Statistics." Accessed September 22, 2017. http://www.iii.org/fact-statistic/pet-statistics

August 27   Winchester, S. *Krakatoa. The Day the World Exploded: August 27, 1883*. New York: Perennial, 2003.

August 28   Thomas, P. *For the Birds: The Life of Roger Tory Peterson*. Honesdale, PA: Calkins Creek, 2011.

August 29   The Friends of Charles Darwin. "John Stevens Henslow." Accessed December 9, 2016. http://friendsofdarwin.com/articles/henslow/

August 30   Institute for Applied Ecology. "Eradication by Mastication." Accessed January 25, 2016. http://appliedeco.org/eradication-by-mastication/

August 31   Thomas, J. "The Role of the Sacred Ibis in Ancient Egypt." janetthomas, March 6, 2013. Accessed May 11, 2016. https://janetthomas.wordpress.com/2013/03/06/the-role-of-the-sacred-ibis-in-ancient-egypt/

September 1   Audubon, J. J. *The Birds of America; from Original Drawings*. London: Published by the Author, 1827–1838.

Peterson, R. T. *Audubon's Birds of America*. New York: Abbeville, 2005.

September 3   Gai Jatra Festival. Accessed May 31, 2016. http://www.weallnepali.com/nepali-festivals/gal-jatra-festival

September 4   Lear, L. *Beatrix Potter: A Life in Nature*. New York: St. Martin's, 2007.

Potter, B. *The Tale of Peter Rabbit*. London: Frederick Warne, 1902.

September 5   Klauber, L. M. *Rattlesnakes: Their Habits, Life Histories, and Influence on Mankind*. Berkeley: University of California Press, 1956.

September 7   Hunter, F. "Lewis & Clark's Prairie Dog: An Odyssey." Frances Hunter's American Heroes Blog, May 25, 2010. Accessed January 25, 2016. https://franceshunter.wordpress.com/2010/05/25/lewis-clarks-prairie-dog-an-odyssey/

September 8   Venzel, S. "Celebrate Reptiles on National Iguana Awareness Day." Wide Open Pets. Accessed September 27, 2017. www.wideopenpets.com/celebrate-reptiles-on-national-iguana-awareness-day

September 9   American Transcendentalism Web. "Nature," by Ralph Waldo Emerson. Accessed January 12, 2017. http://transcendentalism-legacy.tamu.edu/authors/emerson/essays/naturetext.html

September 10   American Museum of Natural History. "In Memoriam Stephen

Jay Gould (1941–2002)." Accessed August 5, 2016. http:// www.amnh.org/science/bios/gould/

September 11   IUCN. "The 100 Most Threatened Species: Are They Price-less or Worthless?" September 11, 2012. Accessed August 29, 2016. http://www.iucn.org/?11022/The-100-most-threatened-species—Are-they-priceless-or-worthless

September 12   Saunders, N. J. *Animal Spirits*. Boston: Little, Brown, 1995.

September 13   Castle, S. "Eagles Trained to Take Down High-Tech Prey: Small Drones." *New York Times*, May 29, 2016.

September 14   Hemming, J. *Tree of Rivers: The Story of the Amazon*. New York: Thames & Hudson, 2008.

von Humboldt, A. *Personal Narrative of Travels to the Equi-noctial Regions of America, During the Years 1799–1804 by Alexander von Humboldt and Aimé Bonpland*. Trans. by T. Ross. Vol. 1–3. London: H. Bohn, 1852.

Wulf, Andrea. *The Invention of Nature: Alexander von Hum-boldt's New World*. New York: Alfred A. Knopf, 2015.

September 15   Galapagos Conservancy. "Giant Tortoises." Accessed January 12, 2017. http://www.galapagos.org/about_galapagos/about-glaapagos/biodiversity/tortoises/

September 16   Guggisberg, C. A. W. *Crocodiles: Their Natural History, Folk-lore and Conservation*. Harrisburg, PA: Stackpole, 1972.

September 17   University of Vermont Extension, Department of Plant and Soil Science. "Asters and Goldenrod: The Mythology." Ac-cessed July 21, 2016. http://www.uvm.edu/pss/ppp/articles/asters.html.

September 19   Ravenstar, D. "Goddess Mama Kilya." Journeying to the Goddess, September 20, 2012. Accessed January 9, 2017. https://journeyingtothegoddess.wordpress.com/2012/09/20/goddess-mama-kilya/

September 20   World Wildlife Fund, "A Stamp to Protect Wildlife." October 15, 2014. Accessed August 3, 2016. http://www.worldwildlife.org/stories/a-stamp-to-protect-wildlife

September 21   Doyle, A. "Syrian War Spurs First Withdrawal from Dooms-day Arctic Seed Vault." Reuters World News. Sep-tember 21, 2015. Accessed February 24, 2017. http://www.reuters.com/article/us-mideast-crisis-seeds-idUSKCN0RL1KA20150921.

September 22   Romain, W. F. "Last Words of Chief Crowfoot." The Ancient Earthworks Project, May 29, 2014. Accessed January 2, 2016. http://ancientearthworksproject.org/1/post/2014/05/last-words-of-chief-crowfoot.html

September 23   Allen, D. *Otter*. London: Reaktion, 2010.

                North Pacific Fur Seal Convention. pribilof.noaa.gov/docu-
                ments/THE_FUR_SEAL_TREATY_OF_1911.pdf

September 24   Reference.com. "What Is the Symbolic Meaning of a Blue-
                bird?" Accessed September 27, 2017. https://www.
                reference.com/world-view/symbolic-meaning-bluebird-
                9e380d390a08db27

September 25   Willis, C. M., S. M. Church, C. M. Guest, et al. "Olfactory De-
                tection of Human Bladder Cancer by Dogs: Proof of Princi-
                ple Study." *British Medical Journal* 329 (2004): 712–714.

September 26   The Beatles Lyrics. "Octopus's Garden." Accessed September
                27, 2017. https://www.azlyrics.com/lyrics/beatles/octopuss-
                garden.html

                The Beatles Ultimate Experience. "Abbey Road." Accessed
                February 28, 2017. http://www.beatlesinterviews.org/
                dba11road.html

September 27   Carson, R. *Silent Spring*. New York: Houghton Mifflin, 1962.

September 28   NASA. "NASA Confirms Evidence That Liquid Water Flows
                on Today's Mars." September 28, 2015. Accessed August 5,
                2016. http://www.nasa.gov/press-release/nasa-confirms-
                evidence-that-liquid-water-flows-on-today-s-mars

September 29   Coitir, N. M. *Ireland's Birds: Myths, Legends and Folklore*. West
                Link Park, Cork, Ireland: Collins, 2015.

September 30   The Galway International Oyster Festival. Accessed May 31,
                2016. http://www.discoveringireland.com/the-galway-
                international-oyster-festival/

October 1   de Waal, F. "What I Learned Tickling Apes." *New York Times*,
                April 10, 2016.

October 3   American Humane Society. "New Study Highlights Educational
                Value of Pets in the Classroom." Accessed January 10, 2017.
                http://www.americanhumane.org/about-us/newsroom/
                news-releases/new-study-highlights-educational-value-of-
                pets-in-the-classroom.html

October 4   Lovejoy, T. E. "Aid Debtor Nations' Ecology." *New York Times,*
                October 4, 1984.

October 5   Dogs for Diabetes (D4D). Accessed August 27, 2016. https://
                dogs4diabetes.com/about-us/

October 7   Armstrong, E. A. *The Life and Lore of the Bird: In Nature, Art,
                Myth, and Literature*. New York: Crown, 1975.

                Poetry Foundation. "The Raven," by Edgar Allan Poe. Accessed
                January 6, 2017. https://poetryfoundation.org/poems-and-
                poets/poems/detail/48860

Rowland, B. *Birds with Human Souls: A Guide to Bird Symbolism*. Knoxville: University of Tennessee Press, 1978.

October 8   Coates, P. *Salmon*. London: Reaktion, 2006.

October 9   Stevenson, J. "Camille Saint-Saëns. Carnival of the Animals, zoological fantasy for 2 pianos & ensemble." AllMusic. Accessed January 12, 2017. http://www.allmusic.com/composition/carnival-of-the-animals-zoological-fantasy-for-2-pianos-ensemble-mc0002658281

Yo-Yo Ma. "The Swan." Saint-Saëns. YouTube.

October 11   US Fish & Wildlife Service. "History of the Bison Herd," Accessed January 12, 2017. https://www.fws.gov/refuge/Wichita_Mountains/wildlife/bison/history.html

October 12   Heaney, S. *Field Work*. New York: Farrar, Straus, and Giroux, 1979.

October 14   Moir, H. M., J. C. Jackson, and J. F. C. Windmill. "Extremely High Frequency Sensitivity in a 'Simple' Ear." *Biology Letters* 2013. Accessed January 2, 2017. http://dx.doi.org/10.1098/rsbl.2013.0241.

October 15   Opsahl, K. "Nylon Replacement?" *Herald Journal*, December 16, 2015.

White, E. B. *Charlotte's Web*. New York: Harper & Brothers, 1952.

October 16   Moose Madness Family Festival. Accessed January 8, 2017. http://www.visitcookcounty.com/plan-your-trip/activities-by-season/fall/moose-madness-family-festival/

October 17   Gorilla100.com. "Discovery." Accessed January 20, 2016. http://www.gorilla100.com/30-Discovery.html

Savage, T. S., and J. Wyman. "Notice on the External Characters and Habits of *Troglodytes gorilla*, a New Species of Orang from the Gaboon River; Osteology of the Same." *Boston Journal of Natural History* 5 (1847): 417–443.

October 18   Defenders of Wildlife. "Wolf Awareness Week." Accessed August 27, 2016. http://www.defenders.org/wolf-awareness-week

October 19   Crump, M. L. *In Search of the Golden Frog*. Chicago: University of Chicago Press, 2000.

Dobzhansky, T. "Evolution in the Tropics." *American Scientist* (1950): 209–22.

Shoemaker-Galloway, J. "Rainforest Day." Accessed August 20, 2016. http://www.holidailys.com/single-post/2015/10/19/Rainforest-Day

October 20   Discovering Lewis & Clark. "Jefferson's Megalonyx." Accessed
              July 24, 2016. http://www.lewis-clark.org/article/2742

October 21   Ceríaco, L. M. P., and M. P. Marques. "Deconstructing a
              Southern Portuguese Monster: The Effects of a Children's
              Story on Children's Perceptions of Geckos." *Herpetological
              Review* 44 (2013): 590–594.

October 22   Wombania. "Wombat Day October 22." Accessed February 24,
              2017. http://www.wombania.com/wombat-day.htm

October 23   Gould, S. J. "Fall in the House of Ussher." *Natural History* 100
              (1991): 12–21.

October 24   NPR History Department. "Hats Off to Women Who Saved
              the Birds." NPR, July 15, 2015. Accessed December
              10, 2016. http://www.npr.org/sections/npr-history-
              dept/2015/07/15/422860307/hats-off-to-women-who-saved-
              the-birds

October 26   Babb, D. "History of the Mule." Accessed September 9, 2016.
              http://www.mulemuseum.org/history-of-the-mule.html

October 27   Northern Plains Potato Growers Association. "Potato Fun
              Facts." Accessed September 9, 2016. http://nppga.org/
              consumers/funfacts.php

October 29   Hewson-Hughes, A. K., A. Colyer, S. J. Simpson, and D.
              Raubenheimer. "Balancing Macronutrient Intake in a
              Mammalian Carnivore: Disentangling the Influences of
              Flavour and Nutrition." *Royal Society of Open Science*,
              June 15, 2016. doi:10.1098/rsos.160081

October 30   Poladian, C. "Tihar Festival in Nepal Celebrates Dogs with
              Garland, Not Skewers [Photos]." *International Business
              Times*, June 26, 2015. Accessed September 2, 2016. http://
              www.ibtimes.com/pulse/tihar-festival-nepal-celebrates-
              dogs-garland-not-skewers-photos-1986154

November 1   From the Journal of Meriwether Lewis. "One Continual Roar."
              Accessed January 3, 2017. http://www.lewis-clark.org/
              article/443

              National Bison Legacy Act. Accessed January 18, 2016. http://
              www.votebison.org/bill

November 2   Manos-Jones, M. *The Spirit of Butterflies: Myth, Magic, and
              Art*. New York: Henry N. Abrams, 2000.

November 3   Sumner, T. "Ancient Jellyfish Died a Strange Death." *Science
              News for Students*, November 14, 2014. Accessed January
              8, 2017. https://student.societyforscience.org/article/
              ancient-jellyfish-died-strange-death

November 4   Crystalinks. "King Tutankhamen's Tomb." Accessed November
              11, 2016. www.crystalinks.com/tutstomb.html

November 5   McVicker, D. "Animism Religion." Prezi, February 12, 2015. Accessed January 2, 2016. https://prezi.com/vry7odfjpde9/animism-religion/

November 6   Netstate. "Texas State Reptile." Accessed January 2, 2016. http://www.netstate.com/states/symb/reptiles/tx_horned_lizard.htm

November 7   Kona Coffee Cultural Festival. Accessed August 18, 2016. http://konacoffeefest.com/about-the-festival/

November 8   Pushkar Camel Fair. Accessed January 7, 2017. http://www.pushkarcamelfair.com/

November 10   Smithsonian Institution. "The Hope Diamond." Accessed December 13, 2016. http://www.si.edu/Encyclopedia_SI/nmnh/hope.htm

November 11   US Department of Veterans Affairs. Office of Public and Intergovernmental Affairs. "In Flanders Fields." Accessed September 11, 2016. http://www.va.gov/opa/vetsday/flanders.asp

November 12   National Chrysanthemum Society, USA. "History of the Chrysanthemum." Accessed January 12, 2017. http://www.mums.org/history-of-the-chrysanthemum/

November 13   Poetry Foundation. "The Rime of the Ancient Mariner," by Samuel Taylor Coleridge. Text of 1834. Accessed January 11, 2017. https://www.poetryfoundation.org/poems-and-poets/poems/detail/43997

November 14   Lyell, C. *Principles of Geology: Being an Attempt to Explain the Former Changes of the Earth's Surface, by Reference to Causes Now in Operation.* 3 vol. London: John Murray, 1830–1833.

                Rampino, M. R. "Reexamining Lyell's Laws." *American Scientist* 105 (2017): 224–231.

November 15   Crump, M. *Eye of Newt and Toe of Frog, Adder's Fork and Lizard's Leg: The Lore and Mythology of Amphibians and Reptiles.* Chicago: University of Chicago Press, 2015.

November 16   Tagore, R. *Fireflies: A Collection of Proverbs, Aphorisms and Maxims.* Wakefield, RI: Asphodel, 2007.

                Western Monarch Count Resource Center. http://www.western-monarchcount.org/

November 17   American Bald Eagle Foundation. "Alaska Bald Eagle Festival." Accessed January 11, 2017. https://baldeagles.org/alaska-bald-eagle-festival/about/

November 18   Ingersoll, E. *Birds in Legend, Fable, and Folklore.* New York: Longmans, Green, 1923.

November 19   Grooms, S. *The Cry of the Sandhill Crane*. Minocqua, WI: NorthWord, 1992.

November 20   Ventana Wildlife Society. "California Condor #799 aka 'Princess.'" Accessed March 13, 2017. http://www.mycondor.org/condorprofiles/condor799.html

November 21   Do-Your-Bit. "World Fisheries Day. 21 November." Accessed August 20, 2016. http://www.gdrc.org/doyourbit/21_11-fisheries-day.html

November 22   Humane Society of the United States. www.humanesociety.org

November 23   BBC. "Religions. Kami." Last updated September 4, 2009. Accessed January 12, 2017. http://www.bbc.co.uk/religion/religions/shinto/beliefs/kami_1.shtml

November 24   Alfred, R. "July 1, 1858: Darwin and Wallace Shift the Paradigm." Wired, July 1, 2011. Accessed March 12, 2017. https://www.wired.com/2011/07/0701darwin-wallace-linnean-society-london/

Darwin, C. *On the Origin of Species by Means of Natural Selection, or The Preservation of Favoured Races in the Struggle for Life*. New York: Avenel, 1979. Originally published 1859.

Darwin, C. *Voyage of the Beagle*. New York: Penguin, 1989. Originally published 1839.

November 25   Humane Society of the United States. "Joseph Wood Krutch: Philosopher of Humaneness." Accessed January 4, 2017. http://www.humanesociety.org/about/history/joseph_wood_krutch.html

Krutch, J. W. "Conservation Is Not Enough." *American Scholar* 23 (1954): 295–305.

November 27   Turks and Caicos Conch Festival. Accessed May 6, 2016. http://turksandcaicostourism.com/turks-and-caicos-conch-festival/

November 28   Monkey Buffet Festival. Accessed December 3, 2015. http://festivalasia.net/festivals/Monkey-Buffet-Festival-2015.html

November 29   Salter, D. *Holy and Noble Beasts: Encounters with Animals in Medieval Literature*. Cambridge: D. S. Brewer, 2001.

November 30   Quinn, B. "Unicorn Lair 'Discovered' in North Korea." *Guardian*, November 30, 2012. Accessed June 23, 2016. https://www.theguardian.com/world/2012/nov/30/unicorn-lair-discovered-north-korea

December 1   HCA.Gilead.org.il. "The Nightingale," by Hans Christian Andersen. Accessed August 11, 2016. http://hca.gilead.org.il/nighting.html

December 2   United States Environmental Protection Agency. epa.gov.

December 3   Thomas, L. *The Lives of a Cell: Notes of a Biology Watcher*. New York: Bantam, 1974.

December 4   World Wildlife Fund. "Wildlife Conservation Day." Accessed September 27, 2017. https://www.worldwildlife.org/stories/wildlife-conservation-day

December 5   Koyeli Tours and Travels. "Hornbill Festival of Nagaland." Accessed January 3, 2016. http://hornbillfestival.co.in/about1.html

December 6   Kahn, T. R., E. La Marca, S. Lötters, J. L. Brown, E. Twomey, and A. Amézquita, eds. *Aposematic Poison Frogs (Dendrobatidae) of the Andean Countries: Bolivia, Colombia, Ecuador, Perú, and Venezuela*. Arlington, VA: Conservation International, 2016.

December 7   Kendall, P. "Holly." Trees for Life. Accessed January 12, 2017. http://www.treesforlife.org.uk/forest/mythology-folklore/holly2/

December 8   Tautin, J. "100 Years of Bird Banding in North America 1902–2002." US Geological Survey. Accessed January 11, 2017. http://www.pwrc.usgs.gov/bbl/homepage/100years.cfm

                USGS. "A Brief History about the Origins of Bird Banding." Accessed January 11. 2017. https://www.pwrc.usgs.gov/BBL/homepage/historyNew.cfm

December 9   Koestler, A. *The Case of the Midwife Toad*. New York: Random House, 1971.

December 10   Animal Equality. "International Animal Rights Day Manifesto 2012." Accessed September 9, 2016. http://www.internationalanimalrightsday.org/

December 11   United Nations. "International Mountain Day." Accessed January 13, 2017. http://www.un.org/en/events/mountainday/background.shtml

December 12   National Day Calendar. "National Poinsettia Day." Accessed September 2, 2016. http://www.nationaldaycalendar.com/national-poinsettia-day-december-12/

                University of Illinois Extension. "Poinsettia Facts." Accessed September 2, 2016. http://extension.illinois.edu/poinsettia/facts.cfm

December 13   NOAA Fisheries. "Chinese River Dolphin/Baiji (*Lipotes vexillifer*)." Updated January 15, 2015. Accessed January 11, 2017. http://www.fisheries.noaa.gov/pr/species/mammals/dolphins/chinese-river-dolphin.html

December 14   Estrada, A., P. A. Garber, A. B. Rylands, et al. "Impending Extinction Crisis of the World's Primates: Why Primates

Matter." *Science Advances* 3 (January 18, 2017). doi:10.1126/sciadv.1600946

Swindler, D. R. *Introduction to the Primates*. Seattle: University of Washington Press, 1998.

Zimmer, C. "Most Primate Species Threatened with Extinction, Scientists Find." *New York Times*, January 18, 2017.

December 15 Coitir, N. M. *Ireland's Birds: Myths, Legends and Folklore*. West Link Park, Cork, Ireland: Collins, 2015.

December 16 Wikisource. "'Goin' Home' lyrics, by William Arms Fisher (1922)." Accessed January 7, 2016. https://en.wikisource.org/wiki/Goin'_Home

December 17 US National Park Service. "1903—The First Flight." Last updated April 14, 2015. Accessed September 9, 2016. https://www.nps.gov/wrbr/learn/historyculture/thefirstflight.htm

December 18 Coitir, N. M. *Ireland's Mammals: Myths, Legends and Folklore*. West Link Park, Cork, Ireland: Collins, 2010.

December 20 Dickens, C. *The Cricket on the Hearth*. London: Bradbury and Evans, 1845.

Xiaomin, X. "Cricket Matches—Chinese Style." *Shanghai Star*, September 4, 2003. Accessed June 16, 2016. http://app1.chinadaily.com.cn/star/2003/0904/fo8-1.html

December 21 Bradford, W. *Of Plymouth Plantation*. Norton Anthology of American Literature. W. Franklin, P. F. Gura, and A. Krupat, eds. Vol. A. New York: Norton, 2007.

Gundersen, J. R. "The Plymouth Colony. The Pilgrims and Plymouth Rock. 1620." Tripod. Accessed August 26, 2016. http://franklaughter.tripod.com/cgi-bin/histprof/misc/plymouth.html

December 22 Newgrange.com. "Newgrange—World Heritage Site." Accessed December 28, 2016. www.newgrange.com/

December 23 Dinofish.com. "'Discovery' of the Coelacanth. The Fish out of Time," site maintained by J. F. Hamlin. Accessed May 31, 2016. www.dinofish.com/discoa.htm

December 24 Martin, L C. *Wildlife Folklore*. Old Saybrook, CT: Globe Pequot, 1994.

Whipp, D. "The History of Santa's Reindeer." Altogether Christmas. Accessed November 18, 2015. www.altogetherchristmas.com/traditions/reindeer.html

December 25 Christmas Carol Lyrics. "The Twelve Days of Christmas." Accessed March 14, 2017. http://www.41051.com/xmaslyrics/twelvedays.html.

December 26   National Audubon Society. "History of the Christmas Bird
               Count." Accessed September 7, 2016. http://www.audubon.
               org/conservation/history-christmas-bird-count

December 27   Great Seal. "Benjamin Franklin on the Rattlesnake as a Symbol
               of America." Accessed August 27, 2016. http://www.great-
               seal.com/symbols/rattlesnake.html

December 28   NOAA Fisheries. "Endangered Species Act (ESA)." Updated
               February 11, 2016. Accessed September 3, 2016. http://
               www.nmfs.noaa.gov/pr/laws/esa/

               US Fish and Wildlife Service. "West Virginia Northern Flying
               Squirrel." Last updated March 4, 2013. Accessed Septem-
               ber 3, 2016. https://www.fws.gov/northeast/newsroom/
               wvnfsq.html

December 29   Elie, P. *Reinventing Bach*. New York: Farrar, Straus, and Gir-
               oux, 2012.

December 30   Gratacap, L. P. "The Development of the American Museum
               of Natural History." *American Museum Journal* 1 (1900–
               1901): 2–4.